DISCOVERING CALCULUS WITH
MAPLE®

Second Edition

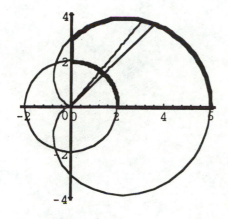

KENT HARRIS
WESTERN ILLINOIS UNIVERSITY

ROBERT J. LOPEZ
ROSE-HULMAN INSTITUTE OF TECHNOLOGY

JOHN WILEY & SONS, INC.
New York   Chichester   Brisbane   Toronto   Singapore

ISBN 0-471-00973-3

Printed in the United States of America

10 9 8 7 6 5 4 3

Printed and bound by Malloy Lithographing, Inc.

Preface

This second edition has been extensively rewritten, reflecting numerous changes. The examples contain greater detail, show many new Maple features, and expand the use of a computer algebra system in solving different types of problems. Most of the exercises are new or have been updated. Each chapter now includes, in addition to a collection of exercises, a set of projects. The exercise sets are generally of no more than moderate difficulty, while the project sets present more challenging problems. Exercises referenced in the text are called "Exercises," whereas projects are called "Problems." The structure is similar to that of most traditional calculus books. In particular, the chapters roughly follow the flow of the fifth edition of Calculus with Analytic Geometry by Howard Anton, and the seventh edition of Calculus One and Several Variables by S. L. Salas and Einar Hille (revised by Garret J. Etgen).

The discussion and examples presented in each chapter are very nearly reproductions of Maple worksheets done on a Macintosh computer using Maple V Release 3. In order to put headers on the pages, and to achieve the correct margins, the worksheets were copied into WordPerfect 3.0. Since WordPerfect also allows Maple graphs to be opened as graphics objects, the liberty was taken to tidy up many of the Maple plots. The look and feel of a typical Maple worksheet, however, was preserved. The Exercises and Projects were written with MathWriter on a Macintosh since Maple itself does not yet allow the level of mathematical typesetting needed for writing those sets of exercises.

This text embodies the experiences of two different styles of using Maple in teaching calculus. At Western Illinois University the Maple activities occur in a lab that is distinct from the classrooms where calculus instruction formally occurs. At Rose-Hulman Institute of Technology the instruction takes place in classrooms equipped with one computer for each student. Our goal was to make this edition useful for either environment. If we have succeeded, we will consider the efforts that generated this book well spent.

To The Student

This material is not intended to replace your calculus text, but rather to help you to understand and apply the ideas of calculus. Before you enter Maple commands, be sure to understand the problem and plan your solution. Some preliminary work with paper, pencil, and intellect will often help you to find an effective solution. **Take the time to do this; you will save time and avoid frustration!**

Maple is fast and accurate in most situations, provided valid commands are entered. However, you should never accept its responses uncritically. Just as with work done by pencil and paper, you should always ask if the results are reasonable. You still need to develop a sense of "correctness" for any type of mathematical output, whether from a computer or from any other source. Because Maple simplifies the task of mathematical experimentation, the real benefit from its use is in finding ways to test and cross-check solutions obtained from it. Can you use a plot to support your answer? Can you arrive at the same answer through a different approach? The insights that result from such pondering will more than repay your efforts at learning Maple syntax.

Maple is a mathematical tool. Like any human activity, knowing how to use the tool is essential for success with it. Like knowing the playbook in football, knowing how to read music in the band, or knowing the vocabulary of a foreign language, it is more than helpful to have mastery of a residue of Maple syntax and functionality. Make a list of commands that are not easy to remember. Add to this list any commands that you will not use regularly. Update your list at least once a week. The insides of book covers are good for this. Learn to use on-line help. You can open a help window at any time and then continue with your work when you have the information you need.

A final observation might be in order for students who were introduced to calculus without the ancillary use of a computer. In more traditional calculus classes, there is more emphasis on the manipulative aspects of calculus. With a computer algebra system at hand, there is less emphasis on manipulations, but increased emphasis on conceptualization and problem solving. This is, in fact, as it should be. The objective of mathematics is mastery of relationships between ideas, not simply mastery of skills that computers now perform better than humans. We urge our students to accept that their abilities to manipulate the expressions of the calculus will never be as highly regarded as their abilities to understand the concepts of calculus and to use this understanding in solving problems.

To The Teacher

The emphasis here is on the calculus, not on Maple, so the examples shown and the problems presented generally do not require complicated Maple solutions. While it is useful to acquire a basic facility with Maple, it is not necessary to be a Maple expert in order to apply Maple to the calculus. Hence, provision should be made for students to become familiar with Maple, but this does not mean that our classes need be given over to lectures in Maple syntax. Both authors of this text believe that the simplicity of Maple syntax, and its closeness to mathematical thinking, shortens the learning time significantly. But it does not vanish, and students need to learn how to implement calculus in Maple if they are to succeed in a Maple-based course.

At Western Illinois University many calculus sections now meet three days per week (instead of four) for two hours (instead of one). The first part of a two hour session takes place in a classroom, with Maple commands and questions an integral part of the classroom discussion. For the second part of the session, the class moves into a lab to work in pairs on homework problems or on a larger lab assignment. There is always good interaction in the lab where the instructor is available primarily to answer questions. Teams comprised of two students work enthusiastically, develop communications skills, and build cooperative working relationships while making efficient use of hardware resources.

At Rose-Hulman Institute of Technology where the computer algebra system is integrated into the calculus course more thoroughly, the students learn the language during the first weeks of the quarter while applying it to mathematics familiar from high school. Thus, graphing, trigonometry, inverse functions, analytic geometry, conics, and solving equations are the domain over which the computer algebra system is first seen. Since this is material that will be covered in some way even in a traditional course, no topics had to be dropped from the syllabus when computers were introduced.

In either environment, it is essential that students do not have unpleasant first experiences while learning calculus with a computer algebra system. Any efforts that prevent initial frustration and confusion are more than helpful; they are essential. Having access to Maple help during the initial phases of the course definitely helps to alleviate difficulties. Having helpers available to circulate in the lab or classroom, looking at computer screens and telling students that their difficulty is nothing more than a missed semicolon, can prevent cases of wretched frustration. In a class of thirty students, one instructor cannot provide everyone the personal attention needed during the first weeks using Maple. Extra help in the lab at this time is very welcome and can provide valuable experience for Maple knowledgeable students. And providing students clear models and scripts of problem solving with Maple is equally necessary for success. That is why this text exists - the authors discovered independently how important it is in a Maple calculus class to have access to usable material.

If you get your students off to an enjoyable start using Maple, they will quickly become interested and competent. **Try to anticipate troubles they will encounter by working through lab assignments under the conditions they will experience.**

If you have IBM or Macintosh machines, you will find that 8MB of RAM allows Release 3 to run smoothly. A coprocessor in these machines will speed up numeric computations, and concomitantly, plotting. Symbolic computations are not noticeably enhanced by a coprocessor. We hope all readers of this text use Maple through a graphical user interface (GUI) such as the Macintosh, the Microsoft Windows, or the XWindows implementation of the larger workstations. Worksheets can be corrected "on screen" in these interfaces, and this text was written with the implicit assumption that all readers would have access to this vastly improved environment for doing Maple. Readers using a teletype interface will find that the command structure of Maple is the same as in a GUI, but that considerably more effort is required to produce a finished document that looks like the worksheets that became this text.

Acknowledgements

We owe thanks to many people: To our colleagues who encouraged our experimentation; to our students who endured our experimentation which sometimes went awry; to our hardware support colleagues who kept our labs running; to the National Science Foundation for equipment financing; to the editors of The Bent of Tau Beta Pi for permitting us to include selected problems from the 'Brain Tickler' section of The Bent; to Jaynalee and Mary Anne for their help in preparing the exercise and projects sets; to Sharon at John Wiley & Sons for her direction and patience; and especially to our wives Mardell and Glenda, whose qualities are too numerous to list.

Kent Harris
Western Illinois University
Robert J. Lopez
Rose-Hulman Institute of Technology

CONTENTS

Chapter 1.   **GETTING STARTED.** Introduction to Maple, ... 1
Different Platforms, Equations, Graphs via Maple.
Examples 1 - 7, Exercises 1 - 18, Projects 1 - 8.

Chapter 2.   **FUNCTIONS AND LIMITS.** Animation, Implicit 19
Functions, Maps, Definition of Limit.
Examples 1 - 6, Exercises 1 - 24, Projects 1 - 17.

Chapter 3.   **DIFFERENTIATION.** Secant and Tangent Lines, ... 39
The Rate of Change, The Derivative.
Examples 1 - 9, Exercises 1 - 28, Projects 1 - 5.

Chapter 4.   **APPLICATIONS OF THE DERIVATIVE.** 59
Extrema, Looping, Newton's Algorithm, Related Rates.
Examples 1 - 7, Exercises 1 - 39, Projects 1 - 9.

Chapter 5.   **INTEGRATION.** Antiderivatives, Area, Riemann Sums, ... 85
The Fundamental Theorem.
Examples 1 - 14, Exercises 1 - 28, Projects 1 - 9.

Chapter 6.   **APPLICATIONS OF THE DEFINITE INTEGRAL.** 107
Constructing Functions, Area, Volume, Arc Length.
Examples 1 - 8, Exercises 1 - 18, Projects 1 - 8.

Chapter 7.   **LOGARITHMIC AND EXPONENTIAL FUNCTIONS** 119
Inverse Functions, Applications.
Examples 1 - 11, Exercises 1 - 22, Projects 1 - 8.

Chapter 8.   **INVERSE TRIGONOMETRIC AND HYPERBOLIC** 137
FUNCTIONS. Plots, Exponential Forms.
Examples 1 - 6, Exercises 1 - 13, Projects 1 - 8.

Chapter 9.   **INTEGRATION - EXACT AND APPROXIMATE.** 147
The Fundamental Theorem, Simpson's Rule, Error Terms.
Examples 1 - 9, Exercises 1 - 17, Projects 1 - 12.

Chapter 10.   **IMPROPER INTEGRALS.** Indeterminates, ... 165
Approximations, The Cauchy Principal Value.
Examples 1 - 7, Exercises 1 - 7, Projects 1 - 6.

Chapter 11.   **SEQUENCES AND INFINITE SERIES.** Plots, ..………… 177
Partial Sums, Taylor Polynomials, Error Terms.
Examples 1 - 6, Exercises 1 - 31, Projects 1 - 24.

Chapter 12.   **ANALYTIC GEOMETRY.** Plots, Fitting a Conic to ………… 197
Points, Transforming Coordinates.
Examples 1 - 2, Exercises 1 -13, Projects 1 - 7.

Chapter 13.   **POLAR COORDINATES AND PARAMETRIC** …………. 207
EQUATIONS. Animation, Parametrization, Generating
Curves, Fonts, Splines.
Examples 1 - 7, Exercises 1 - 16, Projects 1 - 9.

Chapter 14.   **VECTORS AND THREE DIMENSIONAL SPACE** ……. 227
Plot Commands, Surfaces, Curves, Products of Vectors.
Examples 1 - 8, Exercises 1 - 23, Projects 1 - 9.

Chapter 15.   **VECTOR FUNCTIONS.** Unit Tangent Vectors, …………… 241
Curvature.
Examples 1 - 7, Exercises 1 - 15, Projects 1 - 11 .

Chapter 16.   **PARTIAL DERIVATIVES.** Functions of Two ……………… 261
Variables, Contourplots, Tangent Planes, Directional
Derivatives, The Gradient.
Examples 1 - 16, Exercises 1 - 32, Projects 1 - 16.

Chapter 17.   **MULTIPLE INTEGRALS.** Matrixplot, Iterated …………… 289
Integrals, Optimal Coordinates.
Examples 1 - 8, Exercises 1 - 14, Projects 1 - 13.

Chapter 18.   **VECTOR CALCULUS.**   Green's, Stokes', and …………… 307
The Divergence Theorems; Line and Surface Integrals.
Examples 1 - 6, Exercises 1 - 8, Projects 1 - 5.

Chapter 19.   **SECOND-ORDER DIFFERENTIAL EQUATIONS.** …….. 321
Linear Homogeneous and Non-Homogeneous Differential
Equations, Motion of a Spring.
Examples 1 - 10, Exercises 1 - 19, Projects 1 - 12.

References. 341

Index. 342

Chapter 1

Getting Started

Maple Commands for Chapter 1

abs(x);	absolute value of x
assign({y=2,x=3});	has effect of y:=2 and x:=3
cos(x);	cosine of x radians
evalf(expr);	evaluate expression using decimals
evalf(expr,n);	evaluate to n digits
expand(expr);	algebraically expand expression
factor(expr);	factor polynomial
fsolve(f(x)=0,x);	numeric solution of equation
fsolve(f(x)=0,x,a..b);	numeric solution of equation, between a and b
fsolve(f(x)=0,x,complex);	find numerically, all roots of polynomial equation
lhs(equation);	left-hand side of equation
Pi	π (must have capital P)
plot(f(x),x=a..b);	graph f(x) on domain [a, b]
plot({f(x),g(x)},x=a..b);	graph f(x) and g(x) on domain [a, b]
q[1];	select first member of sequence, set, or list q
q := 'q';	"erase" any value assigned to q
simplify(expression);	simplifies expression, more or less
sin(x)	sine of x radians
solve(f(x)=0,x);	symbolically solves equation for x
solve({f(x,y)=0,g(x,y)=0},{x,y});	symbolically solves simultaneous equations
sqrt(x);	square root of x
subs(x=a,f);	replace each x in f with a
tan(x);	tangent of x radians
f:=unapply(expr,var);	convert expression into function f(x)
f:=x^2;	assign to f the expression x^2
f:=x->x^2;	create the function $f(x) = x^2$
f:=<x^2 |x>;	create the function $f(x) = x^2$
f:=proc(x) x^2;end;	create the function $f(x) = x^2$

Introduction

This chapter will use some algebra and trigonometry to introduce you to Maple. Effective use of a tool like Maple resides partly in a familiarity with the command structure since this allows you to direct Maple to do your bidding. It is much like playing a sport or a musical instrument, or speaking a foreign language. If you don't know the basics you are just not going to be able to perform.

Since Maple runs on a variety of computers, we could spend most of this chapter describing hardware platforms instead of illustrating the use of Maple. Fortunately, the "feel" of Maple across the spectrum of platforms is fairly uniform and can be addressed in generic terms.

Thus, when we point out that the basic Maple "workspace" is a Maple worksheet, or when we remind you to end each Maple command with a semicolon (;), we are addressing the user of Maple on any of its platforms.

On a UNIX workstation, a Windows machine, or a Macintosh, for example, the default Maple "prompt" is a greater than sign, ">". On the Macintosh, this can be changed to a bullet (•), and the worksheets comprising this text have, in fact, been written on a Macintosh with a bullet for a prompt. Incidentally, the prompt for Maple on a NeXT computer is a greater than sign that has been stretched to fill the input region from top to bottom.

And that brings us to regions. A Maple worksheet consists of text regions, input regions, and output regions. Changing an input region into a text region is done (via the mouse) from the menu at the top of the Maple worksheet. Most platforms allow this to be done from the keyboard, and users need to read the documentation specific to their own machine. This "interface" documentation can be found "on-line" on some of the platforms (e.g., Windows), but not all (e.g., Macintosh). However, all platforms have an "on-line" Maple help system that is both comprehensive and complete. All support a "Browser" and for Release 3, keyword search.

On the Macintosh, Maple commands get executed if you press the Enter key, not the Return key. On a Windows platform there is only a Return key, so to achieve the effect of a carriage return only, the secret is to press the Shift and Return keys simultaneously.

One other "secret" of interest: Inserting prompts and comment regions, either at the end of the worksheet or in its middle, is again platform dependent. However, on each platform this is done by some form of menu option, with an accompanying keyboard equivalent. For example, on the Macintosh, inserting a new prompt is done by the key combination "Apple-m," while in Windows by "Control-i." Similarly, the techniques for "splitting" a region and "joining" two regions in a worksheet are platform dependent. The soundest advice we can give as authors is to examine the menu options specific to your version of Maple.

But let's not talk the subject to death. Rather, let's begin extending the Maple worksheet of which this introduction is the first region, a text region. Maple does mathematics symbolically, as well as numerically. To understand the implications of this revelation, note the following exchanges with Maple, which will be much more enlightening if the reader follows along in Maple on a computer.

• 5 + 3;

8

Incidentally, spaces are not required in Maple input, but we use them liberally in the text to improve readability. Maple can separate regions with a horizontal line, but these lines have been suppressed in this text.

• 1/2 + 2/3;

$$\frac{7}{6}$$

We hope the last result alerts the reader to the implications of a "symbolic math" program like Maple. Where possible, Maple does mathematics in "closed form," just like you were taught to do "by hand" early in your schooling. If an input contains floating-point (decimal) numbers, Maple will generally return the answer as a floating-point number.

- (2 - 7.3)/5.9;

$$-.8983050847$$

- abs(-5);

$$5$$

- sqrt(4);

$$2$$

- sqrt(5);

$$\sqrt{5}$$

Again, we see what we mean by a "symbolic" program. The square root of 4 is easily returned as 2, but the square root of 5 can only be approximated as a floating-point number. To do that, we introduce the notion of labeling, making liberal use of the letter "q" as a label because it is so easy to find on the keyboard.

- q := sqrt(5);

$$q := \sqrt{5}$$

- evalf(q);

$$2.236067978$$

- evalf(q, 60);

$$2.23606797749978969640917366873127623544061835961152572427090$$

Throughout this text we assume the reader is using a version of Maple with a GUI, that is, a Graphical User Interface. With such an interface, the worksheet provides the editing features that older versions of Maple lacked. Thus, to change an earlier input to Maple, simply use the mouse to insert the cursor (by clicking) at the point of correction or alteration, and insert or delete characters as needed. Hitting the execution key (Return or Enter, as appropriate) will re-execute a Maple command, replacing old output with new.

Since input and output can be selected with the mouse and deleted, it is possible to create unfathomable worksheets, and worksheets with surprises. For example, if the input "q := sqrt(5)" is deleted from the worksheet, the variable q still has the value $\sqrt{5}$. Cutting from the worksheet does not change the memory state. To erase the value assigned to q requires

- q := 'q';

$$q := q$$

In addition to arithmetic, Maple does algebra in a symbolic form.

- q := (x - 2)*(x + 3);

$$q := (x-2)(x+3)$$

- q1 := expand(q);

$$q1 := x^2 + x - 6$$

- q2 := factor(q1);

$$q2 := (x-2)(x+3)$$

We have assigned to q the expression $(x - 2)(x + 3)$. To insert a particular value into the x's in q, we use the **subs** command which performs substitution or replacement.

- subs(x = 0, q);

$$-6$$

To interrogate Maple for the contents of the variable q, do the following.

- q;

$$(x-2)(x+3)$$

At several places in this text we will need functions rather than just expressions. The distinction will be explained more than once since it is a subtlety not easily mastered. There are proponents on each side, some claiming that expressions are easier for students to master, while others feel the added difficulty of mastering Maple's functional notation makes the user hardened and tough. We leave readers to choose for themselves, but an obvious bias toward expressions appears in this text.

To create a function in Maple you have several options.

- f1 := x -> x^2;

$$f1 := x \rightarrow x^2$$

- f2 := <x^2| x>;

$$f2 := \langle x^2 \mid x \rangle$$

- f3 := proc(x) x^2; end;

$$f3 := proc(x) \ x^2 \ end$$

Each notation has created an equivalent function.

- f1(3);
  f2(3);
  f3(3);

$$9$$

$$9$$

$$9$$

There is yet another way to create a function in Maple in the case where the rule for the function has already been entered as a Maple expression.

- f := x^2;

$$f := x^2$$

- f4 := unapply(f, x);

$$f4 := (x \rightarrow x)^2$$

Observe that f(3) does not produce 9. In fact, it produces nonsense. The notation returned for f4 is indeed strange, but nonetheless, f4(3) will be 9.

- f(3);
  f4(3);

$$x(3)^2$$

$$9$$

Maple can even do trigonometry in symbolic form. For example, remembering that π is entered as Pi, we get

- sin(Pi/2);

$$1$$

- tan(Pi/4);

$$1$$

- cos(Pi/4);

$$\frac{1}{2}\sqrt{2}$$

- simplify(sin(x)^2 + cos(x)^2);

$$1$$

We caution the reader once again about Pi. If the letters "pi" are used in a Maple input, Maple will return the symbol π because Maple returns Greek symbols for *all* the Greek alphabet.

However, pi, unlike Pi, is not known to Maple as the ratio between the circumference and the diameter of a circle. This kind of trap is what computer users call a "gotcha" because after the unwary user has been caught in this trap, the computer programmer, gleefully rubbing hands together, grins "got you again!"

One of the most useful features of symbolic mathematics programs like Maple is the facility to produce graphs painlessly.

- plot(f, x = -1..1);

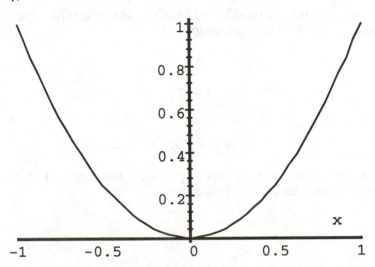

The graph appeared in a separate graphics window, but was copied and pasted into the worksheet by the edit options in the menu. While the graph is in the graphics window, it is "live" and can be altered and manipulated by the "options" available through the options window. After pasting into the worksheet, the graph is inert, except on the NeXT, where the graph can still be resized within the worksheet.

We plotted the expression x^2 when we plotted f. The syntax for plotting a function is slightly different, and it is this difference that suggests the wisdom of learning just one mode of Maple operation and one style of Maple syntax. Thus, the same graph would appear if we entered either plot(f1, -1..1) or plot(f1(x), x = -1..1). Function names like f1 require no "x=" but the "rule" f1(x) behaves like an expression and requires the "x=."

The syntax for plotting several graphs at once, and the syntax for controlling the scale on the vertical axis, are illustrated in the next command.

- plot({f, 2*f}, x = -1..1, 0..1);

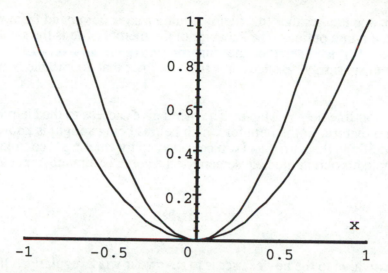

File management on individual platforms differs, but we can make some general remarks that will help users on any platform. First, the user should distinguish between the worksheet files and the "memory" files. The worksheet file is what Maple will store by default when a "save" is issued via the file menu. On a Windows platform this file will automatically receive the ".ms" extension, indicating a worksheet file. If such a file (on a Windows platform) is re-opened at a later time via the "open" file command, the worksheet will be restored on screen, but the memory state for the values of the variables will not be available. The individual commands in the worksheet will have to be re-executed, either one at a time or globally, via the menu option. However, the default filename for a worksheet on a Macintosh will not have the ".ms" extension added, although the file is the worksheet file of any other platform.

There is one last difference to note between the Macintosh and Windows versions. If a pre-existing worksheet is "opened" on a Macintosh via the "open" option on the file menu, the worksheet opens as a "Scratchpad." The Macintosh allows multiple Scratchpads to be open at once, but all such Scratchpads are inert, not directly connected to the Maple kernel. Material from the Scratchpad can be copied and pasted into the active Worksheet, or the curser can be placed in an input region in the Scratchpad and the Enter key pressed. This will cause the input in the Scratchpad to be copied into the Worksheet, and executed in the Worksheet. Thus, by this technique, an old Worksheet can be reconstituted in a new form by selectively choosing portions of it in a Scratchpad format. Portions of this text were, in fact, developed in that manner as incomplete ideas in several Worksheets coalesced into the final forms of some of these chapters. And remember, the Macintosh is the only platform that supports the notion of a Scratchpad.

To save the memory state something special must be done. For example, on a Windows platform one must specifically choose to save the memory state by selecting the appropriate options on the menu. This choice causes Maple to create a file with the same name as the worksheet file, but this time with the ".m" extension. When such a worksheet is re-opened later on, Maple will look to see if a memory file exists, and will load it automatically as it loads the worksheet. In this event, the variables in the worksheet are now all fully defined as if the commands that created them had just been executed.

On the Macintosh, one creates the ".m" file by issuing a **save** command from within the worksheet, not via a menu option. The behavior of the memory file is the same as on other platforms, however. If it is understood that ".m" files are generally very large files, intelligent users will save them sparingly. Besides, it is the ".ms" file that is portable between platforms, not the ".m" file.

A final word about on-line help in Maple: The menu gives access to the Help Browser and to the keyword search feature. If the item for which help is being sought is known to the user, the help screen specific to that command can be called up by entering "?command_name" into Maple. For example, to obtain the help screen for the **solve** command, enter ?solve into Maple.

Examples

The best hint we can give to the novice user is to save your work regularly. Once it is clear to the user that the work done in Maple is "on track," it is probably time to save the work into a file. Periodically thereafter, issue an additional **save** command from the menu to add the new work to the existing file. One never knows when Maple will be sent into a state from which it cannot escape, thereby losing a Worksheet. Smart, indeed, is the user who has taken the precaution to **save, save, save** throughout the session.

Example 1

Find $2/3 + 1/719$ to 15 decimal places.

- evalf(2/3 + 1/719, 15);

$$.668057487250811$$

Example 2

Solve the equation $3 x + 7 = 0$.

- solve(3*x + 7 = 0, x);

$$\frac{-7}{3}$$

Example 3

Solve the equation $x^3 - 8 x + 10 = 0$.

- q := x^3 - 8*x + 10 = 0;

$$q := x^3 - 8x + 10 = 0$$

• q1 := solve(q, x);

$$q1 := -\%2 - \frac{8}{3}\%1, \frac{1}{2}\%2 + \frac{4}{3}\%1 + \frac{1}{2}I\sqrt{3}\left(-\%2 + \frac{8}{3}\%1\right),$$

$$\frac{1}{2}\%2 + \frac{4}{3}\%1 - \frac{1}{2}I\sqrt{3}\left(-\%2 + \frac{8}{3}\%1\right)$$

$$\%1 := \frac{1}{\left(5 + \frac{1}{9}\sqrt{489}\right)^{1/3}}$$

$$\%2 := \left(5 + \frac{1}{9}\sqrt{489}\right)^{1/3}$$

Symbolic solutions can get cumbersome! Maple has introduced two abbreviations, namely, %1 and %2, defined at the end of the sequence of three roots. The commas separating the solutions are significant, as are the I's which stand for $\sqrt{(-1)}$. It appears, therefore, that we have a sequence of three roots, two of which are complex. Obtain a floating-point equivalent of these roots.

• evalf(q1);

$$-3.318628218, 1.659314109 - .5098724705\,I, 1.659314109 + .5098724705\,I$$

We can reference an individual root by the selector notation

• q1[1];

$$-\left(5 + \frac{1}{9}\sqrt{489}\right)^{1/3} - \frac{8}{3}\frac{1}{\left(5 + \frac{1}{9}\sqrt{489}\right)^{1/3}}$$

• evalf(q1[1]);

$$-3.318628218$$

Sometimes it is necessary to solve an equation numerically.

• fsolve(q, x);

$$-3.318628218$$

The floating-point solver obtained just the real root. But it can be instructed to obtain all the roots.

- fsolve(q, x, complex);

$$-3.318628218, 1.659314109 - .5098724701\,I, 1.659314109 + .5098724701\,I$$

Example 4

Plot the graph of the expression on the left side of the equation in Example 3. Note the use of the **lhs** command to obtain the left-hand side of an equation.

- plot(lhs(q), x = -3..3);

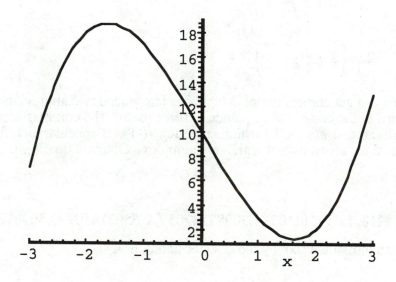

Example 5

Solve the equation |2x - 1| = 9. Draw a graph and use it to approximate the solution, then obtain the exact solution.

- plot({abs(2*x - 1), 9}, x = -5..6);

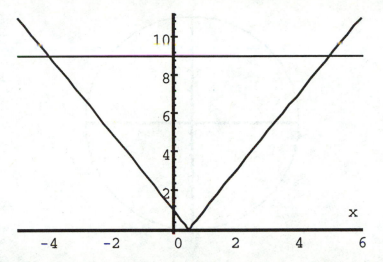

The plot window has a digitizer with which Maple approximates coordinates of the point at the tip of the cursor arrow. If the tip of the cursor arrow is placed at the leftmost intersection of the line y = 9 and y = |2x - 1|, and the mouse button is pressed, Maple will write the coordinates of the point of the tip of the arrow onto the plot window. The accuracy of the digitizer is a function of screen resolution, and hand-eye coordination. On the Macintosh PowerBook with which these lessons were written, the left-most intersection point was given as (-3.969, 8.941) instead of the exact (-4, 9) Maple finds via the **solve** command below.

- solve(abs(2*x - 1) = 9, x);

$$5, -4$$

Example 6

Plot the graph of the relation $x^2 + y^2 = 4$. This is a circle that represents a function $y = y(x)$ implicitly. Hence, we can either use the **implicitplot** command (met later) or we can solve for $y(x)$ explicitly. Here, we will follow the second alternative.

- q := solve(x^2 + y^2 = 4, y);

$$q := \sqrt{-x^2 + 4}, -\sqrt{-x^2 + 4}$$

Again, we have a sequence of two roots. Each root is a formula for a distinct branch of the function $y(x)$ defined implicitly by the equation of the circle. A graph of these two branches drawn on the same axes should give us the circle.

- plot({q}, x = -2..2, scaling = constrained);

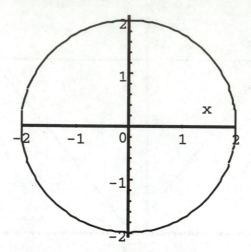

The plot option *scaling*, set to *constrained*, causes both axes to have equal scales. It could be invoked interactively from the options window once the user saw the graph of an ellipse where that of a circle was expected.

Example 7

Solve the system of equations $2x - 4y + 2 = 0$, $5x + 3y = 21$.

- e1 := 2*x - 4*y +2 = 0;
  e2 := 5*x + 3*y = 21;

$$e1 := 2\,x - 4\,y + 2 = 0$$

$$e2 := 5\,x + 3\,y = 21$$

- q := solve({e1, e2}, {x, y});

$$q := \{\,y = 2, x = 3\,\}$$

The task of getting the information about x and y back into Maple is significant, since it means the difference between being easy to use Maple or not. There are two strategies we can use here. One is direct and one is subtle. The direct strategy uses the **assign** command which has the effect of converting the two equations in the set q into the assignments x := 3, and y := 2.

- assign(q);

The **assign** command has only a *null* return, but its effect can be judged via

- x;
  y;

3

2

To "erase" these values from the variables x and y, do the following.

- x := 'x';
  y := 'y';

$$x := x$$

$$y := y$$

The alternative strategy is to use the **subs** command to make a substitution.

- X := subs(q, x);
  Y := subs(q, y);

$$X := 3$$

$$Y := 2$$

The substitution strategy does not make assignments to x or y, and as can be seen, the **subs** command does not have to use all of the equations in the set q.

Exercises

1. Define $f(x) = x \cos x - 3$. Plot $f(x)$ for $0 \le x \le 10$.

   (a) Using Maple's digitizer on the plot window, estimate the largest value of $f(x)$ on your plot.

   (b) "Zoom in" on the highest point by using the command **replot**. (See ? replot.)

   (c) "Zoom in" again until you feel that you can estimate the largest value of $f(x)$ correct to the nearest $\frac{1}{10}$.

2. Repeat Exercise 1 with $f(x) = \tan(1.2\, x - 0.5\, x^2 + 0.83)$ for $0 \le x \le 3$.

3. Find

   (a) $\frac{1}{3} + \frac{1}{7}$ to 10 decimal places.

   (b) $\frac{1}{17} + \frac{1}{177}$ to 15 decimal places.

4. Solve $x^2 + x - 6 = 0$.

5. Solve $x^3 - 2.01\, x^2 - 4.415\, x + 3.2886 = 0$, correct to 15 decimal places.

6. Define $f(x) = 4999\, x^6 + 12124757\, x^4 - 19683\, x^2 - 531441$.

   (a) Plot the graph of $f(x)$.

   (b) Solve $f(x) = 0$, verifying that your roots agree with those in part **(a)**.

7. Define $f(x) = 2\, x \sin x - x^2$.

   (a) Plot the graph of $f(x)$.

   (b) Solve $f(x) = 0$, verifying that your roots agree with those in part **(a)**.

8. Define $f(x) = 3\, x - 4$.

   (a) Plot $f(x)$.

   (b) Solve $f(x) < 0$ and compare your answer with your plot in part **(a)**.

9. Define $f(x) = x^2 - 3x - 9$.

    (a) Plot $f(x)$.

    (b) Solve $f(x) > 0$ and compare your answer with your plot in part (a).

10. Solve $|4x - 5| = |x + 1|$. Can you check your answer graphically?

11. Solve the system of equations $2x + 3y = 13$, $3x - 2y = 0$. Can you check your answer graphically?

12. Plot the graph of $f(x) = \dfrac{2}{3x} + x$

    (a) on $[-5, 5]$.   (c) on $[0, 1]$.

    (b) on $[-1, 1]$.   (d) What is the domain of $f(x)$?

13. Plot the graph of $4x^2 + 9y^2 = 13$. Note that this equation does not give just a single function.

14. Plot the graph of the line that passes through the point $(3, 1)$ and is parallel to the line $3x + 2y = 10$.

15. Plot the upper branch of the parabola $x = \frac{1}{2}y^2 - 1$.

16. A point P moves so that the distance from P to $(3, 0)$ is twice the distance from P to $(-2, 0)$. Graph the locus of P.

17. In the same window, plot the graphs of $y = 3 - x$ and $y = x^2 - 2$. Find any intersection points of the line and the parabola.

18. Solve $4999 x^6 + 1214757 x^4 - 19683 x^2 - 531441 > 0$. Check your answer graphically.

Projects

1. Define $f(x) = 2x^5 - 25x^2 + 7$. Solve each of the following for x. Helpful commands include **fsolve** and **plot**.

    (a) $f(x) = 0$ (b) $f(x) > 5$ (c) $1 \le f(x) \le 2$.

2. Solve each inequality.

(a) $x^5 + 2 x^2 + 7 > 4 x^4 + x$ (b) $\sqrt{x^3 - x + 1} > \sqrt{x^4 + 1}$.

3. A circle in quadrant I is tangent to both axes, and the point $(2, 0)$ is on the circle. Plot the circle.

4. A circle in quadrant II is tangent to both axes, and the point $(-8, 9)$ is on the circle. Plot the circle.

5. Define and display two circles that have the appearance and relation to each other as shown.

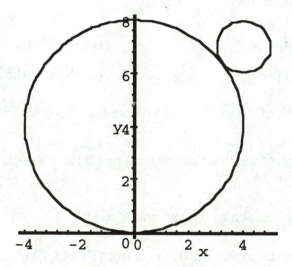

6. Define and display two straight lines as shown.

(a)   (b)

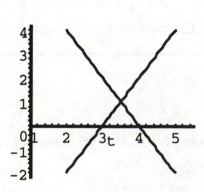   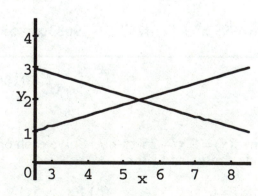

7.   Define the ellipse E by $x^2 + \dfrac{y^2}{9} = 1$. In each of the following problems, plot, in the same window, E and a circle that satisfies the stated requirement (if possible!).

(a) A circle $x^2 + y^2 = r^2$ that does not intersect E

(b) A circle $x^2 + y^2 = r^2$ that intersects E in 2 points

(c) A circle $x^2 + y^2 = r^2$ that intersects E in 4 points

(d) A circle that intersects E in 1 point

(e) A circle that intersects E in 3 points

(f) A circle that intersects E in more than 4 points.

8.   (a) Plot the upper half of the circle $x^2 + y^2 = 25$, and call it C.

(b) In the same window, by trial and error, plot C and a straight-line segment that appears to be tangent to C at the point (4, 3).

(c) In the same window, by trial and error, plot C and a straight-line segment that appears to be normal to C at the point (4, 3). (*Normal* means the line "travels directly away from C.")

Chapter 2

Functions and Limits

New Maple Commands for Chapter 2

animate	see Example 6
for ... do	loop structure; see Example 5
limit(f(x),x=a);	limit of f(x) as x -> a
limit(f(x),x=a,right (or left));	limit of f(x) as x -> a from right (or left)
Limit(f(x),x=a,right (or left));	inert limit, written, not computed
qq:=solve(f(x) = 0, x);	assign to qq the solution of f(x) = 0
seq(k^2,k=1..5);	the sequence 1, 4, 9, 16, 25
value(inert expr);	evaluates inert forms such as Limit
with(plots):	loads plots package
implicitplot(f(x,y),x=a..b,y=c..d);	graphs relation f(x,y) = 0
display([s],insequence=true);	animates sequence of plots, s, into a movie

---

Introduction

The primary objects of the calculus are functions. Hence, it is useful to clarify notions connected with the various features of these "things" called functions. Actually, the concept of function as a mapping or relationship is highly abstract, and most students meet functions via the algebraic rules that define the correspondence that is at the heart of the idea of function.

Maple assists in learning about functions in several ways. First, Maple can manipulate the algebraic formulas by which the rules for functions are stated; second, Maple implements a functional language; and third, Maple draws graphs of functions and relations. From these representations we are assisted in answering questions about functions, questions that include domain, range, behaviors such as increasing and decreasing, and limiting values.

Examples

Example 1

As a first example, consider the function that relates every real number to its square. The rule for this function is expressed by the formula $f(x) = x^2$. In Maple there are at least two ways to represent this function. First, we can represent just the rule, the formula (or expression), x^2.

- f := x^2;

$$f := x^2$$

The assignment operator := attaches the name "f" to the expression x^2. As such, Maple does not recognize the notation f(3). Hence,

19

- f(3);

$$x(3)^2$$

echoes nonsense because f is just a name, a handle, and not the functional f(x). To stick x = 3 into the rule x² we would use the **subs** command that performs substitution.

- subs(x = 3, f);

$$9$$

For those who are ready to deal with functions as mappings, we illustrate Maple's very correct machinery for handling functions as maps. The function whose rule is x² is given in Maple as

- F := x -> x^2;

$$F := x \rightarrow x^2$$

This time, the notation F(3) is recognized by Maple.

- F(3);

$$9$$

The commands and syntax for manipulating the rule of a function and for manipulating the function itself are slightly different. Classroom experience suggests that for the beginner it is probably wiser to treat functions just one way; otherwise the novice will be juggling two sets of commands and their related syntax, and this has been found to be a hindrance to an efficient learning of Maple.

Next, we'll obtain a graph of the function f. Yes, we know that this language is inaccurate, but the more accurate "obtain a graph of the function whose rule is given by the formula f" is just not going to remain unabridged in any setting where students are found.

- plot(f, x = -3..3);

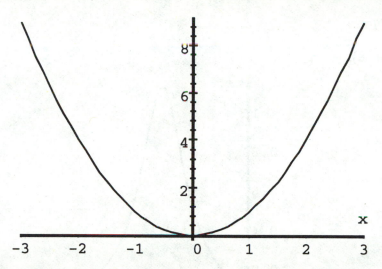

A careful inspection of this plot reveals information about the domain and range of the function. For example, the domain is the set of x's that can legally be substituted into the formula f. In this case, that set is the set of all reals, even though we have elected to graph the function over the interval [-3, 3]. On the other hand, the range is the set of numbers to which the x's map, and on the plot they would be covered by the shadow of the graph if the sun were very far to the right of the origin.

In other instances, values of x that cannot be in the domain of the function are detected by Maple algebraically. For example, consider the function whose rule is

- f1 := 1/x;

$$fl := \frac{1}{x}$$

If we were to try plugging in zero for x, Maple would give an appropriate warning.

- subs(x = 0, f1);

```
Error, division by zero
```

Example 2

A more subtle example of extracting information about a function from its graph is provided by the rational function

- f2 := 10/(10*x^2 - 60*x + 99);

$$f2 := 10 \frac{1}{10 x^2 - 60 x + 99}$$

We begin with a plot.

- plot(f2, x = -4..8);

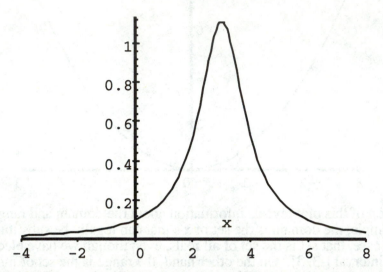

The graph suggests that for x far to the right, or far to the left, the function values are near zero. The domain seems to be all real x's and the range (0, a], where a is the highest value attained at the peak. Issues like the behavior as x moves far to the right or left fall into the arena of limits, and finding a, the largest value in the range, falls into the arena of extreme values. Before exploring either of those items, we look more closely at the question "Does every formula define a function?"

Example 3

Consider the following equation to which we have given the name q.

- q := x*y^2 + y - x = 0;

$$q := x y^2 + y - x = 0$$

This formula is not in the form y = y(x), with the single letter y on the left and only x's on the right. If equation q defines a function of the form y = y(x), then y(x) will be called an implicit function because the x's and y's are intermingled throughout the equation. Sometimes Maple's **implicitplot** command will draw a graph of the solution set of such an equation, and from the graph we can decide if there is a function y = y(x) defined implicitly.

- with(plots):
- implicitplot(q, x = -3..3, y = -3..3);

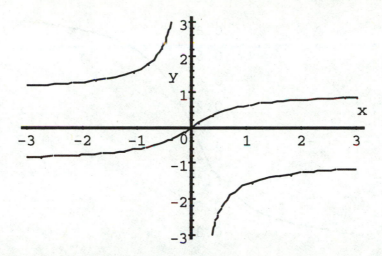

This graph fails the vertical line test wherein a graph is the graph of a function if every vertical line that hits the graph hits it in just one point. Here, a typical vertical line would hit the graph in two points, so this is not the graph of a function. But what if we select "parts" of the graph?

Let's have Maple solve equation q for y = y(x).

- qq := solve(q, y);

$$qq := \frac{1}{2}\frac{-1+\sqrt{1+4\,x^2}}{x}, \frac{1}{2}\frac{-1-\sqrt{1+4\,x^2}}{x}$$

There are two different formulas, and hence, two different functions hidden in the equation q. These functions are called *implicit* functions when hidden in equation q and are called *explicit* when equation q has been solved for y. In this case, let's give separate names to each of the two explicit functions just obtained from equation q.

- y1 := qq[1];
  y2 := qq[2];

$$y1 := \frac{1}{2}\frac{-1+\sqrt{1+4\,x^2}}{x}$$

$$y2 := \frac{1}{2}\frac{-1-\sqrt{1+4\,x^2}}{x}$$

Individual plots of y1 and y2 will pass the vertical line test.

- plot(y1, x = -3..3);

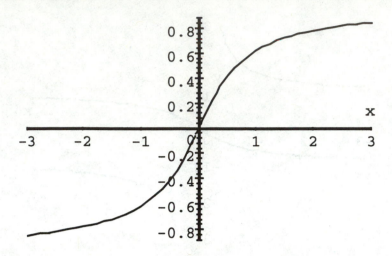

• plot(y2, x = -3..3, y = -10..10);

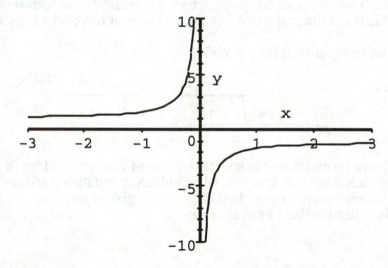

Each formula y1 and y2 contains an x in the denominator, suggesting that perhaps x = 0 should be excluded from the domains. But the graph of y1 seems to be untroubled by this glitch, whereas the graph of y2 has a vertical asymptote at x = 0. The reason for these observations is hidden in the limiting behaviors of y1 and y2, and that is why the concept of limit is so important in the calculus. We'll explore the notion of limit shortly, after we've discussed building new functions from old ones.

Example 4

Given the "old" functions f and g whose rules are

• f := 3*x^2 - 5*x + 17;
  g := 4*x - 9;

$$f := 3\,x^2 - 5\,x + 17$$

$$g := 4\,x - 9$$

we form "new" functions by arithmetic operations between f and g. Thus,

- h1 := f + g;
  h2 := f - g;
  h3 := f * g;
  h4 := f / g;

$$h1 := 3\,x^2 - x + 8$$

$$h2 := 3\,x^2 - 9\,x + 26$$

$$h3 := \left(3\,x^2 - 5\,x + 17\right)\left(4\,x - 9\right)$$

$$h4 := \frac{3\,x^2 - 5\,x + 17}{4\,x - 9}$$

We say that these new functions have been defined "pointwise" since their values at a specific point are determined by the values of f and g at those points. We illustrate this statement.

- f1 := subs(x = 1, f);
  g1 := subs(x = 1, g);
  subs(x = 1, h1), f1+g1;
  subs(x = 1, h2), f1-g1;
  subs(x = 1, h3), f1*g1;
  subs(x = 1, h4), f1/g1;

$$f1 := 15$$

$$g1 := -5$$

$$10, 10$$

$$20, 20$$

$$-75, -75$$

$$-3, -3$$

Composition is another important way of combining two old functions to form a new one. In composition, functions are applied sequentially, one after the other, so that the output of one becomes the input of the other. The composite function is then the totality of the actions of these two operations.

There are two distinct notations used for compositon. The "operational," or formula-based, notation is f(g(x)), whereas the "abstract," or mapping-based, notation is f∘g. Experienced teachers have all discovered that students read this notation as the word "fog" and believe that it properly describes student attitudes toward the abstraction it represents.

Since our "functions" f and g are really expressions, we form the two possible compositions f(g(x)) and g(f(x)) via substitution. The two different composite functions fg = f(g(x)) and gf = g(f(x)) use notation suggestive of the order in which computation and construction are actually done.

- fg := subs(x = g, f);
  gf := subs(x = f, g);

$$fg := 3\,(4\,x-9)^2 - 20\,x + 62$$

$$gf := 12\,x^2 - 20\,x + 59$$

Example 5

Let's return to the question of why the graph of y1 in Example 3 did not have a vertical asymptote at x = 0, whereas the graph of y2 did. Recall the formulas for y1 and y2.

- y1 := (-1 + sqrt(1 + 4*x^2))/(2*x);
  y2 := (-1 - sqrt(1 + 4*x^2))/(2*x);

$$y1 := \frac{1}{2}\frac{-1 + \sqrt{1 + 4\,x^2}}{x}$$

$$y2 := \frac{1}{2}\frac{-1 - \sqrt{1 + 4\,x^2}}{x}$$

When plotting, Maple computes values of the function to be graphed, and then "connects the dots." So, what numbers could Maple have computed prior to plotting y1? What would have happened if Maple had used x = 0 itself as one of the values?

- subs(x = 0, y1);

Error, division by zero

Without doubt, formula y1 is not defined at x = 0. So, Maple's **plot** command did not compute y1 exactly at x = 0, but used values of x near x = 0. What values will formula y1 produce for x's near x = 0? We'll use a for-loop for the repetitious evaluation of the result of substitutions into y1 of ever smaller values of x. The loop ends with **od**, which is clearly *do* spelled backwards.

- for k from 1 to 5 do
      .1^k, evalf(subs(x = .1^k, y1));
      od;

$$.1, .09901951500$$

$$.01, .009999000000$$

$$.001, .001000000000$$

$$.0001, .0001000000000$$

$$.00001, 0$$

As the value of x became smaller, the value of y1 also became smaller. In fact, by the time x was 1/100000 Maple computed y1 to be zero in 10-digit arithmetic. But is that reasonable? Can Maple compute more accurately?

- for k from 1 to 5 do
      .1^k, evalf(subs(x = .1^k, y1), 20);
      od;

$$.1, .099019513592784830000$$

$$.01, .0099999000019995000150000$$

$$.001, .00099999990000000200000000$$

$$.0001, .000099999999990000000000000$$

$$.00001, .0000100000000000000000000$$

The **evalf** command takes a second argument, an integer, that specifies the number of digits the floating-point evaluation is to contain. At this extended precision we see that when x = 1/100000, a more reasonable value for y1 is 1/100000. There are two immediate conclusions to be drawn. First, it appears that for x small enough, y1 behaves like the function y = x. That observation seems consistent with the graph of y1, but the verification of this "local linearity" is at the very heart of the calculus and will be explored in subsequent chapters.

Second, since y1 gets small as x gets small, Maple was not remiss when it connected the dots in the graph of y1. The single point (0, 0) should be omitted from the graph. Even if Maple had done this, it would have been impossible to detect the missing point on the graph. Hence, we would likely say that the limiting value of y1 as x gets near zero is itself zero. In fact, this idea of **limit** is a built-in function in Maple.

- limit(y1, x = 0);

$$0$$

Since our numerical experiment above used only positive numbers, x marched in toward zero *from the right*. This directionality can be reproduced in Maple. In fact, the **limit** command has an inert form by which Maple writes the desired limit without evaluating it.

- q1 := Limit(y1, x = 0, right);

$$ql := \lim_{x \to 0+} \frac{1}{2} \frac{-1 + \sqrt{1 + 4x^2}}{x}$$

The "+" following the "0" under the word "lim" indicates that the limit was taken by progressing through values to the right of $x = 0$. This inert limit can be evaluated via

- value(q1);

$$0$$

With these new tools we can examine the different behavior of y2.

- limit(y2, x = 0);

$$undefined$$

Maple has detected a feature the graph of y2 demonstrates. At a vertical asymptote where the branches of the graph go in opposite directions, the limit does not exist. Maple captures that truth in the word *undefined*. Limits from the left and right will show this to be the proper outcome.

- limit(y2, x = 0, right);
  limit(y2, x = 0, left);

$$-\infty$$

$$\infty$$

Example 6

As a last example, we illustrate the formal definition of limit. We'll use the rational function

- f := (x^2 - 1)/(x - 1);

$$f := \frac{x^2 - 1}{x - 1}$$

for which $x = 1$ is not a valid input (i.e., $x = 1$ is not in the domain of this function). In fact,

- subs(x = 1, f);

```
Error, division by zero
```

However, the limiting value of values generated near x = 1 will be

- limit(f, x = 1);

$$2$$

Indeed, the fraction in f simplifies to

- simplify(f);

$$x + 1$$

at all points where x ≠ 1. It is from this reduced form of the fraction that we determine the limit to be 2 when x gets near 1.

The formal definition of limit requires we show that the values of f are indeed as close to 2 as desired, provided the values of x are near enough to 1. Well, how close to 2 should the values of f be shown to be? Suppose we want these values to be in the interval [2 - e, 2 + e] for e > 0. What is the corresponding restriction on x that accomplishes this? Notice that this question makes the resulting x-interval depend on e. In fact, the x's at the ends of the corresponding x-interval are numbers d = d(e), which can be found algebraically as follows.

- d1 := solve(f = 2 - e, x);
  d2 := solve(f = 2 + e, x);

$$d1 := -e + 1$$

$$d2 := 1 + e$$

Thus, to keep f within e of 2, we need to be sure x is within e of 1. In fact, the values of the function f at the endpoints of the x-interval [1 - e, 1 + e] are

- f1 := simplify(subs(x = d1, f));
  f2 := simplify(subs(x = d2, f));

$$f1 := 2 - e$$

$$f2 := 2 + e$$

The relationship between the x-interval around x = 1 and the interval around y = 2 can be explored graphically. It can even be animated so that we can see dynamically how changing the tightness of the interval around y = 2 affects the permitted x-interval around x = 1.

For this animation we create functions that generate vertical-line segments from the x-axis to the graph of f.

- seg1 := t->[[1 - t, subs(e = t, f1)], [1 - t, 0]];
  seg2 := t->[[1 + t, subs(e = t, f2)], [1 + t, 0]];

$$seg1 := t \rightarrow [[1 - t, \text{subs}(e = t, f1)], [1 - t, 0]]$$

$$seg2 := t \rightarrow [[1 + t, \text{subs}(e = t, f2)], [1 + t, 0]]$$

Next, we create a function whose output is a graph with horizontal lines bounding the interval about y = 2, and vertical lines bounding the x-interval about x = 1.

- F := e->plot({f, 2+e, 2-e, seg1(e), [[1+e,2+e], [1+e,0]]}, x = -1..2);

$$F := e \rightarrow \text{plot}(\{f, 2 + e, 2 - e, seg1(e), [[1 + e, 2 + e], [1 + e, 0]]\}, x = -1 .. 2)$$

We test this function F for some small value of e.

- F(.1);

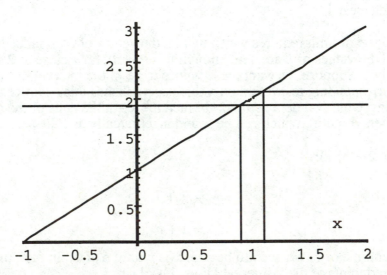

What we'd next like to do is generate a sequence of such pictures, each with a different value of the tolerance e. If we can show these images in succession, we'll have a movie that animates the dependence of the x-interval on the choice of the tolerance interval around f = 2.

- s := seq(F(k/10), k = 1..10):

The **seq** command has created a sequence of plot structures that will become animated by the **display** command in the plots package. Loading a Maple package requires the with command. The colon (:) at the end of a command suppresses output to the screen. The brackets around s change the sequence into a list, and setting the **insequence** flag to *true* creates the animation.

- with(plots):
- display([s], insequence = true);

Paper is clearly not the appropriate medium for demonstrating an animation. The reader is encouraged to execute the Maple code given above to see the effects live.

Exercises

Some of your Maple solutions will be more efficient if you first do some thinking with paper and pencil in use. Also, try to identify any problems that can be worked without the help of Maple.

1. Find a five decimal-place approximation to πx^2 when

    (a) $x = 2$ **(b)** $x = 3.5$

2. Find a solution to $x \sin x + 2 = 0$

    (a) between π and $\dfrac{3\pi}{2}$ **(b)** between $\dfrac{3\pi}{2}$ and 2π.

3. Plot the graph of the function in Exercise 2 to verify that the roots just found are feasible.

4. Find the largest possible domain of the function $f(x) = \dfrac{4}{x^3 - 2.01\, x^2 - 4.415\, x + 3.2886}$.

5. Find the largest possible domain of the function $f(x) = \dfrac{x}{\sqrt{2.5 + 1.2\, x - x^2}}$.

6. Find the approximate range of $f(x) = \dfrac{40}{11\, x^2 - 42\, x + 48}$ (assuming that the largest possible domain is used).

7. Define $f(x) = \dfrac{\sqrt{3 + 2\, x^2}}{x + 5}$.

    (a) Find $f(100.0)$ and then $f(2000.0)$. **(b)** Find $\lim\limits_{x \to +\infty} f(x)$.

8. Define $g(h) = \dfrac{\sin h}{1 - \cos h}$.

    (a) Find $g(0.2)$, $g(0.05)$, and $g(0.001)$. **(b)** Find $\lim\limits_{h \to 0} g(h)$.

9. Define $g(x) = \dfrac{x^2 + x - 2}{x^2 - 4\, x + 3}$.

(a) Plot g on [0, 5] and then on [0.95, 1.05].

(b) Can you find g(1)?

(c) Can you find $\lim_{x \to 1} g(x)$?

10. Define $f(x) = x^2$ and $g(x) = \sin x$.

   (a) Define s(x) as the sum of f(x) and g(x). Plot f(x), g(x), and s(x) in the same window.

   (b) Define p(x) as the product of f(x) and g(x). Plot f(x), g(x), and p(x) in the same window.

   (c) Define c(x) as the composition f(g(x)). Plot f(x), g(x), and c(x) in the same window.

   (d) From the plots, can you see how f(x) and g(x) are used to form each of s(x), p(x), and c(x)?

11. The relation $2y^2 + xy - 5 = 0$ implicitly defines functions of the form y = y(x).

   (a) Find the functions and plot their graphs in the same window.

   (b) Plot these functions using the command **implicitplot**. (Look in the "plots" package.)

12. Determine whether the relation $y^3 + 3xy^2 + 3x^2y + x^3 - x = 0$ defines any functions of the form y = y(x). If so, what are the domains of these functions?

13. For $f(x) = \dfrac{\sqrt{x^2 + 5}}{2x - 1}$ and $g(x) = \dfrac{\sqrt{x + 1}}{x - 2}$, find a decimal value of

   (a) (f + g)(5)   (b) (f / g)(3)   (c) (f ∘ g)(4).

14. Consider the family of functions defined by $y = Ax^2 + Bx + 4$, for any values of A and B, with A ≠ 0.

   (a) What can you say about each member of this family?

   (b) Find the member of the family which passes through the points (4, 4) and (5, 2).

   (c) Find the member which passes through the points (4, 4) and (6, 7).

   (d) Plot the functions from parts (b) and (c) in the same window for -1 ≤ x ≤ 7.

15. Define $f(x) = \sqrt{1 + x}$ and $g(x) = \dfrac{1}{x^2 - 1}$.

   (a) Find $(f \circ g)(x)$ and $(g \circ f)(x)$.

   (b) For $x > 1.2$, plot, in the same window, the graphs of $(f \circ g)(x)$ and $(g \circ f)(x)$.

16. Find an equation of the straight line that passes through the points of intersection of the circles $x^2 + y^2 = 25$ and $(x - 4)^2 + (y - 3)^2 = 16$.

17. Given that $f(x) = x^3$ and $(f \circ g)(x) = (x - 1)^3$, find $g(x)$.

18. For $f(x) = \dfrac{1}{x}$, find a decimal value of

   (a) $f(0.01)$ (b) $f(0.0001)$

   (c) What do the answers to parts (a) and (b) suggest about $\lim\limits_{x \to 0^+} f(x)$?

19. For $f(x) = \dfrac{1}{x}$, find a few specific values of $f(x)$ that will suggest the value of $\lim\limits_{x \to 0^-} f(x)$.

20. For $f(x) = \dfrac{1}{x}$, find

   (a) $\lim\limits_{x \to 0^+} f(x)$ (c) $\lim\limits_{x \to +\infty} f(x)$

   (b) $\lim\limits_{x \to 0^-} f(x)$ (d) $\lim\limits_{x \to -\infty} f(x)$.

21. Define $f(x) = \dfrac{x - 4}{\sqrt{x - 4}}$ (Assume that $x > 4$).

   (a) Find (if possible) a decimal value of $f(4.5)$; $f(4.1)$; $f(4.01)$; $f(4)$.

   (b) What does part (a) suggest about $\lim\limits_{x \to 4^+} f(x)$?

   (c) Find $\lim\limits_{x \to 4^+} f(x)$. (What does Maple say about $\lim\limits_{x \to 4^+} f(x)$?)

   (d) Can you use Maple to simplify $f(x)$ and then find the limit by inspection?

22. Define $f(x) = \dfrac{x^3 - 1}{x - 1}$ (for $x \neq 1$).

(a) Find (if possible) a decimal value of $f(.5)$; $f(.9)$; $f(.99)$; $f(1)$.

(b) What does part (a) suggest about $\lim\limits_{x \to 1} f(x)$?

(c) Find $\lim\limits_{x \to 1} f(x)$.

(d) Can you use Maple to simplify $f(x)$ and then find $\lim\limits_{x \to 1} f(x)$ directly?

23. Find each limit if it exists. (Can you find these limits without using Maple?)

(a) $\lim\limits_{x \to 1} \dfrac{1 - x^2}{2 - \sqrt{x^2 + 3}}$

(c) $\lim\limits_{x \to 2} \dfrac{\sqrt{1 + \sqrt{2 + x}} - \sqrt{3}}{x - 2}$.

(b) $\lim\limits_{x \to 0} \dfrac{\sqrt{x + 1} - 1}{x}$

24. For $f(x) = \sin x$, find

(a) $\lim\limits_{x \to 0} f(x)$

(b) $\lim\limits_{x \to 0} \dfrac{f(x)}{x}$.

Projects

1.   Plot the graph of $f(x)$ and then the graph of $|f(x)|$:

(a) $f(x) = x$

(c) $f(x) = x^2 - 6x + 5$

(b) $f(x) = 2x + 1$

(d) $f(x) = \cos x$.

2.   For each $f(x)$, define $g(x)$ such that the graph of $g(x)$ is the graph of $f(x)$ shifted up by two units. Plot $f(x)$ and $g(x)$ in the same window.

(a) $f(x) = x^2$

(c) $f(x) = \sin x + \dfrac{1}{2x}$ for $1 \le x \le 4\pi$.

(b) $f(x) = \cos x$

3.   Repeat Problem 2, shifting to the right two units instead.

4. A function f(x) is said to be even if f(x) = f(-x) for each x in the domain. A function f(x) is said to be odd if f(x) = -f(-x) for each x in the domain. Decide whether each function is even, odd or neither, by plotting on an appropriate interval.

(a) f(x) = x^2 **(e)** f(x) = cos x

(b) f(x) = x^3 **(f)** f(x) = sin x

(c) f(x) = x^2 + x^3 **(g)** f(x) = cos x + x^2

(d) f(x) = 1 + x^2 - x^4 **(h)** f(x) = sin x - 2 x + x^3.

What can you say about the sum of two (or more) even functions? of two (or more) odd functions?

5. The following plots were generated by combining sin px, cos qx,and x$^{m/n}$ for different values of p, q, m, and n. Define and plot functions that yield a similar appearance. It will help to recognize some plots as a combination of even and/or odd functions.

(a)   **(b)**

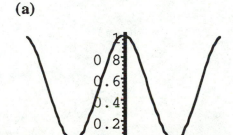   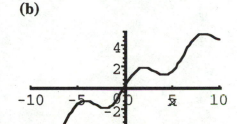

(c)

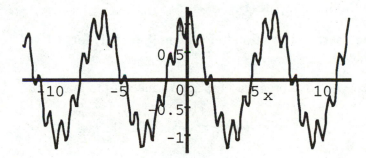

6.   Plot sin x and sin 3x in the same window for $0 \le x \le 2\pi$.
      (a) What can you say about the period and amplitude of each of these functions?

      (b) If n is a positive integer, how would the graphs of sin x and sin nx differ?

      (c) On the interval $[0, 2\pi]$, where are the x-intercepts of sin x? of sin nx?

7.   (a) Plot sin x and sin $0.1x^2$ in the same window for $-10 \le x \le 10$.

      (b) Find the period and amplitude of each of these functions.

8.   Why is it incorrect to refer to $x^2 + y^2 = 9$ as a function? Can you plot the graph defined
      by this relation? There are two functions lurking here. After you have extracted and
      plotted them, open the help file for the command **implicitplot** and rework the problem.

9   In this project you are asked to modify the graph of f(x) = sin x by choosing appropriate
      values for a, b, and c in g(x) = a sin(bx) + c. Each part asks that you choose a, b, and c
      such that g(x) has its first positive root at x = xr, and that the specified maximum value of
      g(x) occurs at x = xm. Check each of your answers by generating a plot.

	xr (First positive root)	xm(Abscissa of max)	g(xm)(Max value)
(a)	π	$\frac{\pi}{2}$	1
(b)	$\frac{\pi}{2}$	$\frac{\pi}{4}$	1
(c)	$\frac{\pi}{2}$	$\frac{\pi}{4}$	0.3
(d)	$\frac{\pi}{3}$	$\frac{\pi}{6}$	1.5
(e)	2.2	1.1	0.25
(f)	2.5	1	2

10.   (a) Find a second solution to Problem 9, part (f).

      (b) Find a range for c such that there is a solution to Problem 9, part (f) for
          each value of c in that range.

11.   Plot each function (or pair) after you have predicted the appearance of the graph(s).

      (a) sin x, $0 \le x \le 8$ (e) $\frac{1}{10}x^2 + \sin 10x, 0 \le x \le 2\pi$

      (b) { sin x, sin (x + 0.5)}, $0 \le x \le 2\pi$ (f) $\frac{1}{10}x^2 + \frac{1}{4} \sin 10x, 0 \le x \le 8$

(c) { $\sin x, \sin 2x$ }, $0 \le x \le 2\pi$ **(g)** $\sin x + \frac{1}{6} \sin 15x$, $0 \le x \le 8$.

(d) { $\sin x, \frac{1}{2} \sin x$ }, $0 \le x \le 2\pi$

Which term causes the rapid oscillation along the smoother path? What causes the larger or smaller amplitude along the curve?

12. Assume that the pitcher's mound on a baseball diamond is a segment of a sphere. If the diameter of the mound is 18 feet at its base and if the volume of the mound is 0.02% of the total volume of the sphere from which it was sliced, what is the height of the mound in inches to two decimal places?

                                                              - Larry L. Lesser (<u>The Bent</u>)

13. The graph of $f(x) = (x - 2)^2 + 1$ is a parabola with vertex at the point (2, 1).

   (a) Define $f_t(x)$ to be the parabola defined by $f(x)$ with its vertex translated to the point (x_1, y_1). Write the appropriate Maple instructions which will display the graphs of $f(x)$ and $f_t(x)$ in the same window.

   (b) Define $f_r(x)$ to be the parabola defined by $f_t(x)$ rotated clockwise through $\frac{\pi}{2}$ radians.Write the appropriate Maple instructions which will display the graphs of $f(x)$ and $f_r(x)$ in the same window.

14. Define $f(x) = 3x - 1$. Since $\lim_{x \to 2} f(x) = 5$, it is true that for any $\varepsilon > 0$, there exists a number $\delta > 0$, such that if $0 < |x - 2| < \delta$, then $|f(x) - 5| < \varepsilon$. For $\varepsilon = 0.4$, find the largest value of δ for which $0 < |x - 2| < \delta$ guarantees that $|f(x) - 5| < \varepsilon$. (Finding the δ is equivalent to finding how close x must be to 2.) *Suggestion:* Plot $f(x)$ for $1.5 \le x \le 2.5$, then rewrite the two inequalities without using absolute values. Look at the plot and understand why a solution can be found by solving "$f(x) - 5 = 0.4$."

15. Repeat Problem 14 for $f(x) = x^3 - 3x - 4$, the limit as x approaches 1.5, with $\varepsilon = 0.05$.

16. Repeat Problem 14 for $f(x) = \ln x + \tan x$, the limit as x approaches 1.5, with $\varepsilon = 0.02$. In this problem you should work with the limit value to six decimal places.

17.   Define $f(x) = x^2 - 6x + 11$, and observe that $\lim_{x \to 4} f(x) = 3$. For each $\varepsilon > 0$, there is some $\delta > 0$ such that $3 - \varepsilon < f(x) < 3 + \varepsilon$ whenever $4 - \delta < x < 4 + \delta$. Create an animation which shows the relationship between ε and δ as ε ranges from 0.05 to 0.5. (See Example 6, but notice that this problem does not possess the symmetry of that example.)

Chapter 3

Differentiation

New Maple Commands for Chapter 3

allvalues(RootOf(expr), 'd');	extract all distinct algebraic roots from **RootOf**
arctan(x);	arctangent of x in radians
D(f);	derivative of *function* f, returns a *function*
diff(expr,x);	derivative of expression with respect to x
diff(f(x),x);	derivative of f(x) with respect to x
diff(f,x,x);	second derivative of f with respect to x
diff(f,x$3);	third derivative of f with respect to x
RootOf	algebraic numbers which are roots of equation
with(student);	load student package
showtangent(f(x),x=c,a..b);	graph f(x) and a tangent line at x = c

---

Introduction

The derivative is the fundamental object of the differential calculus. It is both glib and profound to describe the derivative as the rate of change of a function. Understanding what this phrase means is not a trivial undertaking. Each discipline that uses calculus tries to explain "rate of change" in its own language. In mathematics, the explanation centers on the geometry of secant and tangent lines. In physics, the word "velocity" is used. In economics, "marginal cost" would be uttered. But whatever the language, the essence of the derivative is the limiting value of the ratio of changes in the dependent variable to changes in the independent variable. The exercises of Chapter 3 are designed to illuminate this connection in a geometric setting that reflects the language of mathematics.

Examples

Example 1

If $f(x) = x^2 - x + 3$, have Maple compute f(2).

• F := x^2 - x + 3;

$$F := x^2 - x + 3$$

Note that we have entered an expression whose name (or tag) is the letter "F". In this case, Maple will not understand the notation F(2); For example, try

• F(2);

$$x(2)^2 - x(2) + 3$$

The justaposition of x and the symbol (2) indicates that Maple does not think F can be used functionally. To get the value $x = 2$ into the expression F, use the **subs** command.

• subs(x = 2, F);

$$5$$

Maple also supports functional notation so that the symbols f(2) can be correctly interpreted. There is some overhead, in the form of special syntax, connected with this approach.

First, simply convert the existing expression F to the function f(x) by using Maple's **unapply** command.

• f := unapply(F, x);

$$f := x \rightarrow x^2 - x + 3$$

Maple now understands the meaning of f(2).

• f(2);

$$5$$

Finally, we illustrate how Maple could have been instructed initially to create the function f(x), without first having to know about the expression F. Let's call the function created g(x) for clarity.

• g := x -> x^2 - x + 3;

$$g := x \rightarrow x^2 - x + 3$$

Again, Maple will understand the notation g(2).

• g(2);

$$5$$

In the ensuing examples we will again see some of the differences between working with expressions and working with functions.

Example 2

On the same set of axes, plot the graphs of $f = \sin(x)$ and $g = \cos(2x)$ for x in the interval [0, 7].

• plot({sin(x), cos(2*x)}, x = 0..7);

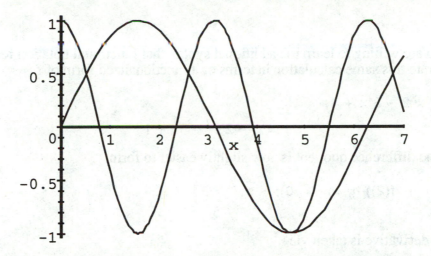

Example 3

By using the definition, compute, at x = 2, the slope of the line tangent to f(x) if f(x) is given by

• f := 2*x^2 - 3*x + 1;

$$f := 2\,x^2 - 3\,x + 1$$

Notice that f represents a Maple expression and we will need to use the **subs** command to evaluate it at particular values. In particular, dq, the difference quotient, is formed via

• dq := (subs(x = 2 + h, f) - subs(x = 2, f))/h;

$$dq := \frac{2\,(2+h)^2 - 8 - 3\,h}{h}$$

The limit of this difference quotient is the limit of the slopes of secant lines. Hence, the limit of dq as h approaches zero will be the slope of the tangent line.

• limit(dq, h = 0);

$$5$$

Notice that this value is actually the derivative evaluated at x = 2. To see this, compute f'(2) via the Maple steps

• f1 := diff(f, x);

$$f1 := 4\,x - 3$$

• subs(x = 2, f1);

$$5$$

For those who are willing to learn the additional syntax that functional notation requires, we next demonstrate this same calculation in terms of a functionalized form of f.

• f := x -> 2*x^2 - 3*x + 1;

$$f := x \rightarrow 2\,x^2 - 3\,x + 1$$

The limit of the difference quotient is now slightly easier to form.

• limit((f(2+h) - f(2))/h, h = 0);

$$5$$

However, the derivative is taken via

• diff(f(x), x);

$$4\,x - 3$$

We started with the function f(x), but the derivative is returned as the expression 4 x - 3. To evaluate this derivative at x = 2, we would need the **subs** command used above. Hence, we have fallen back into the realm of expressions.

To operate totally in the realm of functions, we have to learn even more Maple syntax. Here, we introduce the Maple **D** operator for differentiating functions as functions.

• f1 := D(f);

$$f1 := x \rightarrow 4\,x - 3$$

Notice that the **D** operator returns f1 now as a function from which we can obtain f'(2) via

• f1(2);

$$5$$

Example 4

Plot the graphs of f(x) and the line tangent to f(x) at the point (3, 3) if f(x) is

• f := x -> x^2 - x - 3;

$$f := x \rightarrow x^2 - x - 3$$

The slope of the tangent line is found by computing the limiting slope of the secant lines passing through (3, 3). The difference quotient dq contains the slopes of all such secant lines.

• dq := (f(3+h) - f(3))/h;

$$dq := \frac{(3+h)^2 - 9 - h}{h}$$

The limit of this fraction as h approaches zero is the slope of the tangent line.

• s := limit(dq, h = 0);

$$s := 5$$

This is the same number we get if we substitute x = 3 into the derivative of f(x).

• subs(x = 3, diff(f(x), x));

$$5$$

The tangent line itself, in the form y = s(x - 3) + 3, is given in Maple as

• g := s*(x - 3) + 3;

$$g := 5\,x - 12$$

The required graph is then produced by

• plot({f(x), g}, x = 0..5);

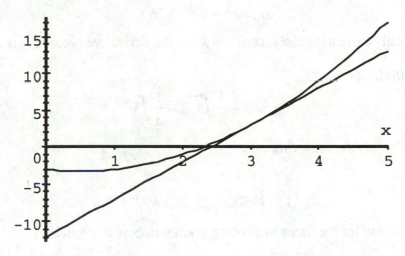

An alternative solution uses the built-in **showtangent** command found in the student package. We use a colon as a terminator to suppress the output of the **with** command that loads the student package.

• with(student):
• showtangent(f(x), x = 3);

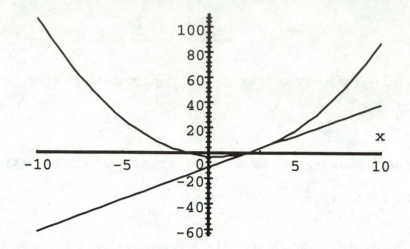

Example 5

If the motion of a particle is given by the formula x(t) below, at what times does the particle stop?

• xt := t^3 - 9*t^2 + 26*t;

$$xt := t^3 - 9\,t^2 + 26\,t$$

Motion stops, if only instantaneously, at times when the derivative (velocity) is zero.

• q := solve(diff(xt, t), t);

$$q := 3 + \frac{1}{3}\sqrt{3}, 3 - \frac{1}{3}\sqrt{3}$$

As floating-point numbers, these times are

• evalf(q);

$$3.577350269, 2.422649731$$

An alternative solution for the times as floating-point numbers is given by

• fsolve(diff(xt, t) = 0, t);

$$2.422649731, 3.577350269$$

Example 6

Find a first quadrant point of intersection of the curves defined implicitly by formulas c1 and c2 below. Graph these curves for visual insight into their relationship.

- c1 := 4*x^2 + 9*y^2 = 36;
  c2 := y^2 = 2*x - 2;

$$c1 := 4\,x^2 + 9\,y^2 = 36$$

$$c2 := y^2 = 2\,x - 2$$

The simplest way to obtain the required graph is via the **implicitplot** command in the plots package.

- with(plots):
- implicitplot({c1, c2}, x = -5..5, y = -5..5);

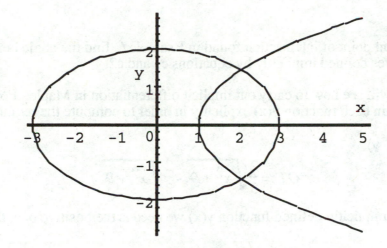

There are at most two points of intersection and only one in the first quadrant.

- q := solve({c1, c2}, {x,y});

$$q := \left\{ y = \text{RootOf}(\_Z^4 + 13\,\_Z^2 - 32), x = \frac{1}{2}\text{RootOf}(\_Z^4 + 13\,\_Z^2 - 32)^2 + 1 \right\}$$

When Maple represents a solution as a **RootOf** expression, a call to the **allvalues** command usually provides an answer in radicals. The second argument can be any letter, but the use of the letter "d" is a reminder that now, only the "distinct" combinations of root-pairs will be delivered.

- q1 := allvalues(q, d);

$$q1 := \left\{ y = \frac{1}{2}\sqrt{-26 + 6\sqrt{33}}, x = -\frac{9}{4} + \frac{3}{4}\sqrt{33} \right\}, \left\{ x = -\frac{9}{4} + \frac{3}{4}\sqrt{33}, y = -\frac{1}{2}\sqrt{-26 + 6\sqrt{33}} \right\},$$

$$\left\{ y = \frac{1}{2}\sqrt{-26 - 6\sqrt{33}}, x = -\frac{9}{4} - \frac{3}{4}\sqrt{33} \right\}, \left\{ x = -\frac{9}{4} - \frac{3}{4}\sqrt{33}, y = -\frac{1}{2}\sqrt{-26 - 6\sqrt{33}} \right\}$$

There are four solutions, but only two can be real. To see this, convert these exact solutions to their floating-point equivalents.

• q2 := evalf(q1);

$$q2 := \{\, y = 1.454937789, x = 2.058421985\,\}, \{\, x = 2.058421985, y = -1.454937789\,\},$$
$$\{\, y = 3.888038576\,I, x = -6.558421985\,\}, \{\, x = -6.558421985, y = -3.888038576\,I\,\}$$

Only the first two solutions are real, and just the first one is in quadrant one. This point can be referenced in Maple as q1[1] or q2[1].

Example 7

At the first quadrant point of intersection found in Example 6, find the angle between the lines tangent to the curves defined implicitly by equations c1 and c2.

In Example 8 we will see how to carry out implicit differentiation in Maple. For the case at hand, we will obtain each function y(x) explicitly in order to compute the required derivatives.

• Q1 := solve(c1, y);

$$Q1 := \frac{2}{3}\sqrt{-x^2 + 9}, -\frac{2}{3}\sqrt{-x^2 + 9}$$

The branch of the implicitly defined function y(x) we need is the positive one, the first one listed in Q1.

• y1 := Q1[1];

$$y1 := \frac{2}{3}\sqrt{-x^2 + 9}$$

Similarly, we obtain y2.

• Q2 := solve(c2, y);

$$Q2 := \sqrt{2\,x - 2}, -\sqrt{2\,x - 2}$$

• y2 := Q2[1];

$$y2 := \sqrt{2\,x - 2}$$

The slopes of the tangent lines at the intersection point are therefore the derivatives evaluated at the intersection.

• m1 := simplify(subs(q1[1], diff(y1, x)));

$$m1 := -\frac{2}{3}\frac{-3+\sqrt{33}}{\sqrt{-26+6\sqrt{33}}}$$

- m2 := simplify(subs(q1[1], diff(y2, x)));

$$m2 := 2\frac{1}{\sqrt{-26+6\sqrt{33}}}$$

Since a derivative is the tangent of the angle the tangent line makes with the x-axis, the angle itself is arctan(y'). Hence

- angle1 := arctan(m1);
  angle2 := arctan(m2);

$$angle1 := -\arctan\left(\frac{2}{3}\frac{-3+\sqrt{33}}{\sqrt{-26+6\sqrt{33}}}\right)$$

$$angle2 := \arctan\left(2\frac{1}{\sqrt{-26+6\sqrt{33}}}\right)$$

- angle2_1 := evalf(angle2 - angle1);

$$angle2\_1 := 1.163483428$$

Since a radian is roughly 57 degrees, this angle is slightly more than 60 degrees. In fact, the number of degrees in the angle would be

- evalf(convert(angle2_1, degrees));

$$66.66268993\ degrees$$

Example 8

By implicit differentiation, compute the derivative of y(x) defined by curve c1 from Example 6.

Conceptually, if we think of each letter "y" in c1 as y(x), then we will see how to proceed with the differentiation. So, begin by replacing each y with y(x) in c1.

- C1 := subs(y = y(x), c1);

$$C1 := 4\,x^2 + 9\,y(x)^2 = 36$$

Maple is case sensitive, and we need to distinguish between c1 and C1. Differentiate C1 with respect to x.

- CC1 := diff(C1, x);

$$CC1 := 8\,x + 18\,y(x)\left(\frac{\partial}{\partial x}y(x)\right) = 0$$

Since we are looking for the derivative y'(x), solve equation CC1 for the unknown derivative, using its full name diff(y(x), x).

• yp := solve(CC1, diff(y(x), x));

$$yp := -\frac{4}{9}\frac{x}{y(x)}$$

To evaluate this derivative at a point such as (x, y) = (a, b), we will have to deal with the y(x) in yp. We can either replace y(x) with the simple letter "y" or we can proceed cautiously with a substitution. Direct substitution will need both the set braces AND the use of y(x) as shown below.

• subs({x = a, y(x) = b}, yp);

$$-\frac{4}{9}\frac{a}{b}$$

The alternative is to replace y(x) with y in yp *before* evaluating at (a, b).

• yp1 := subs(y(x) = y, yp);

$$yp1 := -\frac{4}{9}\frac{x}{y}$$

• subs(x = a, y = b, yp1);

$$-\frac{4}{9}\frac{a}{b}$$

Example 9

If y(x) is defined implicitly by equation c3 below, obtain a graph of y(x), state its domain and range, and find the slope of the tangent line at the point (1, 0).

• c3 := x^2 * y^2 + 8*x*y = 8*y;

$$c3 := x^2\,y^2 + 8\,x\,y = 8\,y$$

First, an implicit plot.

• implicitplot(c3, x = -6..6, y = -4..4, axes = boxed);

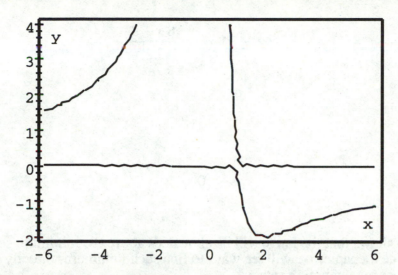

The fact that Maple struggles to complete this graph is an indication of its difficulty. The axes have been removed so that we can see the line y = 0 as part of the graph. However, Maple connects the segments incorrectly because too few points were used in the analysis.

If we solve for y(x) explicitly, we can more clearly see the behavior of y(x).

- C3 := solve(c3, y);

$$C3 := 0, -\frac{8\,x - 8}{x^2}$$

This result shows that the line y = 0 is part of the graph of y(x) and that the y-axis is a vertical asymptote for y(x). Consequently, the domain consists of all real numbers (x = 0 is allowed in formula c3, producing y = 0 as a companion.) The range consists of real numbers r satisfying

- q := C3[2] = r;

$$q := -\frac{8\,x - 8}{x^2} = r$$

Since both limits

- L1 := Limit(C3[2], x = -infinity);
  L2 := Limit(C3[2], x = infinity);

$$L1 := \lim_{x \to (-\infty)} -\frac{8\,x - 8}{x^2}$$

$$L2 := \lim_{x \to \infty} -\frac{8x - 8}{x^2}$$

evaluate to zero via

- value(L1);
  value(L2);

$$0$$

$$0$$

the graph of y(x) shows that the range will be all real x's above the minimum seen in the fourth quadrant. In the next chapter we will see ways to find such minimums directly via the calculus but here we can use an algebraic strategy.

The range will consist of the numbers r, determined by equation q, as x varies through all real numbers. Hence, if we ask "how will an x depend on the companion r," we are led to solve equation q for x = x(r).

- xr := solve(q, x);

$$xr := -\frac{1}{2}\frac{8 + 4\sqrt{4 + 2r}}{r}, -\frac{1}{2}\frac{8 - 4\sqrt{4 + 2r}}{r}$$

Since x must remain real, r cannot be less than -2, so the range will be all reals in the interval [-2, infinity).

Exercises

1. Define $f(x) = x^3 - 3x^2 + 2$ and $g(x) = -\frac{1}{4}x + \frac{1}{2}$.
    Plot $f(x)$ and $g(x)$ in the same window for

    (a) $x \in [0, 1]$ **(b)** $x \in [0.9, 0.92]$.

2. Define $f(x) = x^3 - 3x^2 + 2$.

    (a) At $x = \frac{1}{2}$, find $h(x)$, the line tangent to $f(x)$.

    (b) Plot $f(x)$ and $h(x)$ in the same window for $x \in [0, 1]$.

    (c) Plot $f(x)$ and $h(x)$ in the same window for $x \in [0.4, 0.6]$.

3. Plot $f(x) = |x|$ on $[-0.1, 0.1]$. Can you find a line tangent to $f(x)$ at

    (a) $x = 0.05$ **(b)** $x = 0$.

4. If $f(x) = 2(x - 2)^2$, write appropriate Maple instructions for finding the slope of a secant line on $f(x)$ for the endpoints a, b. Apply these steps when

    (a) $a = 2, b = 4$ **(d)** $a = 3, b = 5$

    (b) $a = 2, b = 3$ **(e)** $a = 3, b = 4$

    (c) $a = 2, b = 2.2$ **(f)** $a = 3, b = 3.1$.

5. Define $f(x) = 2(x - 2)^2$.

    (a) Find the slope of a secant line to $f(x)$ at the points where $x = a$ and $x = a + h$.

    (b) Use your work in part **(a)** and the limit to find the slope of the line tangent to $f(x)$ at the point where $x = 2$; where $x = 3$.

6. Use the limit of the slope of a secant line to find the slope of the tangent line to $f(x) = x^2 + 3x - 2$, at the point where

    (a) $x = 1$ **(c)** $x = 3$

    (b) $x = 2$ **(d)** $x = a$.

You can save time by working part (**d**) first, and then using your results for parts (**a**), (**b**) and (**c**).

To avoid repetition in Exercises 7, 8, and 9, note that the solutions depend on $\frac{(f(a+h) - f(a))}{h}$, or on a limit of this expression.

7. A particle moves along a coordinate axis such that its directed distance from the origin is $s = \sqrt{3t + 2}$ at time = t seconds. Find the average velocity (as a decimal value) of the particle over each time interval:

    (**a**) between t = 1 and t = 4   (**c**) between t = 1 and t = 1.1.

    (**b**) between t = 1 and t = 3

8. A particle moves along a coordinate axis such that its directed distance from the origin is $s = \sqrt{3t + 2}$ at time = t seconds. Use the definition of the derivative to find the instantaneous velocity of the particle at the end of

    (**a**) 1 second   (**c**) 3 seconds

    (**b**) 2 seconds   (**d**) t seconds.

9. A particle is traveling according to $s = t^2 - 3t + 1$. Write Maple instructions for finding the average velocity of the particle between the times t = 5 and t = t_0. Apply your Maple code when

    (**a**) $t_0 = 8$ (**b**) $t_0 = 6$ (**c**) $t_0 = 5.1$.

    (**d**) From your results in parts (**a**), (**b**), and (**c**), can you give a reasonable estimate of the velocity at t = 5?

    (**e**) Find the velocity at time t = 5.

10. Define $f(x) = \dfrac{4}{x\sqrt{x^2 + 7}}$.

    (**a**) Plot f (x) and the secant line through (2, f(2)) and (3, f(3)) in a window where $1 \le x \le 4$.

    (**b**) Write Maple instructions that give the slope of the secant line through (2, f(2)) and (x, f(x)). Apply this Maple code for x = 3, 2.5, 2.1 and 2.01.

    (**c**) As x approaches 2 from the right, what happens to the values of these slopes?

(d) Choose values of x approaching 2 from the left. What is the limit of the values of these slopes now?

(e) Plot f(x), the tangent line at x = 2 and two of the secant lines all in the same window. What happens to the secant lines?

11. Use the definition of the derivative to differentiate

(a) $x^2 - x$ **(d)** sin (cos x)

(b) $\dfrac{x}{x - 1}$ **(e)** cos (x sin x).

(c) sin x

12. A particle moves along a coordinate axis such that its directed distance from the origin is $f(t) = 2\,t^4 - 26\,t^3 + 5\,t^2 + 400\,t - 10$. At what times does the particle stop?

13. **(a)** Plot the graph of the serpentine $y = \dfrac{x}{x^2 + 1}$.

(b) Find the slope of each tangent line to this curve at the points where $|y| = \frac{1}{2}$.

(c) At what points is the slope of the tangent line equal to 0?

14. Find the slope of the line tangent to the graph of the function at the indicated point:

(a) $f(x) = \dfrac{x^2 - x}{x + 1}$ at x = 0

(b) $f(x) = \tan x$ at $x = \dfrac{Pi}{4}$

(c) $f(x) = \exp(x)$ at x = 1.5

(d) $f(x) = \frac{1}{4}x^4 - \frac{5}{2}x^3 + \frac{1}{2}x^2 + 40\,x$ at x = 0.5.

15. Can you find a point on the graph of $f(x) = \frac{1}{4}x^4 - \frac{5}{2}x^3 + \frac{1}{2}x^2 + 40\,x$ at which the tangent line is parallel to the line x - 2 y = 10?

16. **(a)** For $y = x^8 - 2\,x^7 + 3\,x^5 - x^2 + 4\,x$, find $\dfrac{d^4 y}{dx^4}$.

(b) For $y = \sin (2x)$, find $\dfrac{d^{10}y}{dx^{10}}$.

17. Find a cubic polynomial that has a relative minimum at $(-2, 0)$ and a relative maximum at $(3, 4)$.

18. Find a polynomial that has a relative maximum at $(-1, 1)$, a relative minimum at $(2, -3)$, and passes through the origin.

19. Plot the graph of

    (a) $y = \sin x$ on $[-\pi, \pi]$ **(c)** $y = \sin 2x$ on $[0, 2\pi]$.

    (b) $y = \sin x$ on $[0, 4\pi]$

20. Find an equation of the tangent line to the graph of $y = \sin x$ at the point on the curve where

    (a) $x = 0$ **(b)** $x = \dfrac{\pi}{4}$ **(c)** $x = \dfrac{\pi}{2}$.

21. In an ideal system, if a weight is attached to a spring and pulled down 6 inches from its rest position and then released, its motion is described by $y = 6 \cos t$ at any time $t \geq 0$.

    (a) How fast is the weight moving at time $t = 0$? $t = 1$ second? $t = 5$ seconds?

    (b) At what times is the velocity $= 0$?

22. Find each of the following if x represents angular measure in radians:

    (a) $\lim\limits_{x \to 0} \dfrac{\cos x - 1}{x}$ **(b)** $\lim\limits_{x \to 0} \dfrac{\sin x}{x}$ **(c)** $\dfrac{d}{dx} \sin x$.

23. Find the answers to parts **(a)** and **(b)** of Exercise 22 if x represents angular measure in degrees.

24. Use the diff operator to find the derivative of

    (a) $\sin x$ **(b)** $\sin (\cos x)$ **(c)** $\cos (x \sin x)$.

25. **(a)** Plot the graph of the four-cusped hypocycloid $x^{2/3} + y^{2/3} = 1$.

    (b) Find the slope of the tangent line to this graph at the point in the first quadrant where $x = \dfrac{1}{2}$; where $x = 0.9$.

26. **(a)** Plot the graphs of $y^2 = 4\,x^3$ and $2\,x^2 + 3\,y^2 = 14$ on the same coordinate system.

(b) Find a point of intersection P of the two curves in the first quadrant.

(c) Show that the tangent lines at P are perpendicular.

27. Find $\dfrac{dy}{dx}$:

(a) $x^2 + y^2 = 1$ **(c)** $4\,x^2y - 3\,y = x^3 - 1.$

(b) $x^2y^2 + x\,y^3 - x^3 = 1$

28. Each of the following equations defines y implicitly as a function (or more than one function) of x.

(a) Find the domain of y(x).

(b) Find the range of y(x).

(c) Find the derivative $y'(x)$.
    i) $x^2y - y = x^2 + 1$
    ii) $y^3 + 4\,y = x^2$
    iii) $y^3 + 4\,y = x^3$
    iv) $(2\,x - y)^2 = 4\,x\,y$
    v) $x^{2/3} + y^{2/3} = 1$

Projects

1. Obtain the derivative of each the following functions by applying the definition. As an aside during each calculation, find the slope of the secant line through the points (2, f(2)) and (2.5, f(2.5)).

(a) $f(x) = C\,x^3$ **(b)** $f(x) = C\,x^{-4}$

(c) $f(x) = \sin(x)$

Can you recognize where the two special limits, $\displaystyle\lim_{x \to 0} \frac{\sin(h)}{h}$, and

$\displaystyle\lim_{x \to 0} \frac{(\cos(h) - 1)}{h}$, are used?

(**d**) $f(x) = x^4$ and $g(x) = x^3 + 5x$.

In this case, let $q(x) = f(g(x + h)) - f(g(x))$. When you have your answer, check it by applying the Chain Rule using paper and pencil.

2.   Define f: $= x \wedge 3 - 3 * x \wedge 2 + 5$;

(**a**) Print the slope of each secant line on f(x) which passes through (3, f(3)) and (3 + h, f (3 + h)) for h = 1, 0.9, 0.8, ... , 0.1. A loop controlled by a "for" works efficiently:

> for h from 1 by -0.1 to 0.1 do
> ...
> ...
> od;

(**b**) Repeat part (**a**) for h = 0.1, 0.09, 0.08, ... , 0.01.

(**c**) Evaluate the limit of the secant expression to find the slope of the line tangent to f(x) at the point [3, f(3)].

(**d**) Plot f(x), the tangent line and two or three of these secant lines in the same window. *Hint:* In the student package, there is a command **slope**. Also, points can be named; for example, p1: = [3, f(3)].

3.   Useful commands: **plot** or **implicitplot** (with (plots))

(**a**) Plot the portion of the graph of the four-cusped hypocycloid $x^{2/3} + y^{2/3} = 1$ which lies in the first quadrant.

(**b**) Find that tangent line to the graph at the point in the first quadrant where $x = \frac{1}{8}$.
Find the length of that tangent line which is in the first quadrant.

(**c**) Repeat part (**b**) when the point of tangency is taken where $x = \frac{1}{2\sqrt{2}}$.

(**d**) Repeat part (**b**) when the point of tangency is taken where x = c (with 0 < c < 1). Show that this length equals that found in parts (**b**) and (**c**).

4.   (**a**) For $f(x) = x^3 - 3x^2$ and g(x) = x f(x - 1), show that the roots of g'(x) form a geometric progression.

(**b**) For $f(x) = 4(4x^4 + 5) + 25x(x^2 - 2x - 1)$ and g(x) = x f(x), identify the pattern in a sequence of the roots of g'(x).

(**c**) Show that for P(x) = (x - a) (x - [a + d]) (x - [a + 2 d]) (x - [a + 3 d]) the roots of

P'(x) form an arithmetic progression.

(d) Show that if P(x) is a fifth degree polynomial whose roots form an arithmetic progression, then the roots of P''(x) also form an arithmetic progression.

5.   Define $f(x) = \sin(x)$. Estimate $f'(0.5)$ using

(a) $df1 := \dfrac{f(0.5 + h) - f(0.5 - h)}{h}$ for $h = \dfrac{1}{2.}, \dfrac{1}{2.^2}, \cdots, \dfrac{1}{2.^{10}}$

(b) $df2 := \dfrac{f(0.5 + h) - f(0.5 - h)}{2 * h}$ for $h = \dfrac{1}{2.}, \dfrac{1}{2.^2}, \cdots, \dfrac{1}{2.^{10}}$

<u>Suggestion</u>: Use the "for loop" shown below to compare the accuracy of each approximation.

```
cosval: = cos (0.5):
h: = 1:
for k from 1 to 10 do
        h: = h/2. :
        df1: =
        df2: =
        er1: = abs (df1  - cosval):
        er2: = abs (df2  - cosval):
        print (h, er1, er2)
    od;
```

(For better control of your printed output, see ?printf.)

Chapter 4

Applications of the Derivative

New Maple Commands for Chapter 4

evalc(imaginary expr);	causes evaluation of complex quantities
isolate(eqn, var);	solves eqn for the unknown variable
readlib(isolate);	makes isolate available
	isolate also contained in student package
print(a,b);	prints values of a and b
with(plots);	loads plots package
display([p1,p2]);	puts two separately created plots into one
textplot([2,3,`xxx`]);	writes "xxx" at point (2, 3) on a graph

Introduction

The derivative as the limiting slope of secant lines, and hence as the slope of the tangent line, answers an interesting question about curves: Do curves, like lines, have slope? In addition to this facet, it is important that we see the utility of the derivative concept; else why invest time and energy in learning the calculus? In this chapter we explore some applications of the derivative.

Examples

Example 1

Find all points on the graph of

- f := 3*x^4 - 10*x^3 - 29*x^2 + 70*x + 6;

$$f := 3\,x^4 - 10\,x^3 - 29\,x^2 + 70\,x + 6$$

where the tangent line is parallel to the line

- g := 3*x - 2*y = 5;

$$g := 3\,x - 2\,y = 5$$

The first step should be a plot of f and the line given implicitly in g. We can either put the line into the form y = y(x) or we can graph g implicitly. In this case it is probably easier to make the line explicit. In the explicit form, the slope of the line will be obvious.

- q := solve(g, y);

$$q := \frac{3}{2}x - \frac{5}{2}$$

• plot({f, q}, x = -2..4);

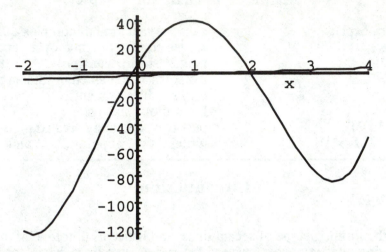

The scale of the function f is distorting the picture, but careful inspection suggests there are three points where the slope of the tangent line might be parallel to a line with slope 3/2. Since the slope of the tangent line is the derivative of f, perhaps we should be looking to see where the graph of f'(x) has the value 3/2.

• fp := diff(f, x);

$$fp := 12\,x^3 - 30\,x^2 - 58\,x + 70$$

• plot({fp, 3/2}, x = -2..4);

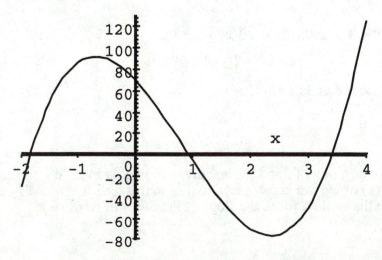

Again, the scale of the function f'(x) is distorting the picture. The horizontal line y = 3/2 is so close to the x-axis that it appears to be part of a thickened axis. But again, the evidence seems to suggest three points where the slope is 3/2. Perhaps it is time to solve for the three points where f'(x) = 3/2.

- q1 := solve(fp = 3/2, x);

$$q1 := \%1^{1/3} + \frac{83}{36}\frac{1}{\%1^{1/3}} + \frac{5}{6},$$

$$-\frac{1}{2}\%1^{1/3} - \frac{83}{72}\frac{1}{\%1^{1/3}} + \frac{5}{6} + \frac{1}{2}I\sqrt{3}\left(\%1^{1/3} - \frac{83}{36}\frac{1}{\%1^{1/3}}\right),$$

$$-\frac{1}{2}\%1^{1/3} - \frac{83}{72}\frac{1}{\%1^{1/3}} + \frac{5}{6} - \frac{1}{2}I\sqrt{3}\left(\%1^{1/3} - \frac{83}{36}\frac{1}{\%1^{1/3}}\right)$$

$$\%1 := -\frac{113}{432} + \frac{1}{432}I\sqrt{2274379}$$

The equation fp = 3/2 is a cubic, and the complexity of the exact roots is disappointing. In fact, each root seems to contain the imaginary number I. But cubics cannot have three complex roots, so the expressions in q1 must simplify. We use the **evalc** command to force Maple to do some complex arithmetic on the three solutions contained in q1.

- r1 := simplify(evalc(q1[1]));
  r2 := simplify(evalc(q1[2]));
  r3 := simplify(evalc(q1[3]));

$$r1 := \frac{1}{3}\sqrt{83}\cos\left(-\frac{1}{3}\arctan\left(\frac{1}{113}\sqrt{2274379}\right) + \frac{1}{3}\pi\right) + \frac{5}{6}$$

$$r2 := -\frac{1}{6}\sqrt{83}\cos\left(-\frac{1}{3}\arctan\left(\frac{1}{113}\sqrt{2274379}\right) + \frac{1}{3}\pi\right) + \frac{5}{6}$$
$$-\frac{1}{6}\sqrt{3}\sqrt{83}\sin\left(-\frac{1}{3}\arctan\left(\frac{1}{113}\sqrt{2274379}\right) + \frac{1}{3}\pi\right)$$

$$r3 := -\frac{1}{6}\sqrt{83}\cos\left(-\frac{1}{3}\arctan\left(\frac{1}{113}\sqrt{2274379}\right) + \frac{1}{3}\pi\right) + \frac{5}{6}$$
$$+\frac{1}{6}\sqrt{3}\sqrt{83}\sin\left(-\frac{1}{3}\arctan\left(\frac{1}{113}\sqrt{2274379}\right) + \frac{1}{3}\pi\right)$$

All the expressions now appear to be real, but they are still more complicated than a simple integer. As floating-point numbers, these are

- evalf([r1, r2, r3]);

$$[3.424622482, -1.833654407, .9090319233]$$

Incidentally, we could have instructed Maple to obtain these roots as decimal approximations immediately via

- fsolve(fp = 3/2, x);

$$-1.833654407, .9090319245, 3.424622482$$

The y-coordinates corresponding to these x-values are

- for k from 1 to 3 do
    y.k := evalf(subs(x = r.k, f));
    od;

$$y1 := -83.38991997$$

$$y2 := -124.2944720$$

$$y3 := 40.20522551$$

The Maple loop illustrates *concatenation*, the joining of the symbol "y" with the numbers 1, 2, and 3, to form new symbols y1, y2, and y3. Think of this as dynamically subscripting y.

Should we need the actual equation of the tangent line at, say, the rightmost point, we could have Maple generate the formula y - y1 = 3/2(x - r1).

Example 2

Find values of a, b, and c that cause the parabola $y = a x^2 + b x + c$ to pass through the point (1, 1) and to have a relative maximum at (3, 4).

Since there are three constants (a, b, and c) to find, we will need three equations that involve these unknowns. For example, the condition that the parabola pass through the point (1, 1) generates one equation. The condition that the parabola have a maximum at (3, 4) implies the parabola must pass through that point too, giving rise to a second equation. Finally, if there is to be a maximum when x = 3, then that is a third condition leading to the final equation.

- q2 := a*x^2 + b*x + c;

$$q2 := a x^2 + b x + c$$

- e1 := subs(x = 1, q2) = 1;

$$e1 := a + b + c = 1$$

- e2 := subs(x = 3, q2) = 4;

$$e2 := 9 a + 3 b + c = 4$$

- e3 := subs(x = 3, diff(q2, x)) = 0;

$$e3 := 6\,a + b = 0$$

- q3 := solve({e1, e2, e3}, {a, b, c});

$$q3 := \left\{ c = \frac{-11}{4}, a = \frac{-3}{4}, b = \frac{9}{2} \right\}$$

We can obtain the parabola itself and draw its graph with the following commands.

- q4 := subs(q3, q2);

$$q4 := -\frac{3}{4}x^2 + \frac{9}{2}x - \frac{11}{4}$$

- plot(q4, x = 0..6);

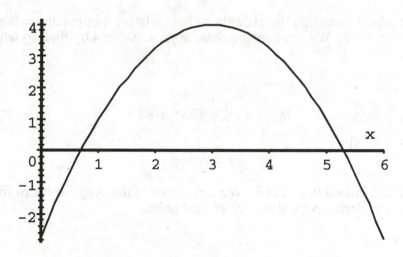

Example 3

Find the relative extrema of the function $y = 2\,x^4 - (40/3)\,x^3 + 24\,x^2 + 2$; determine where it is decreasing and where it is concave up.

It would be very helpful to start with a graph of y.

- q5 := 2*x^4 - (40/3)*x^3 + 24*x^2 + 2;

$$q5 := 2\,x^4 - \frac{40}{3}x^3 + 24\,x^2 + 2$$

- plot(q5, x = -1..4);

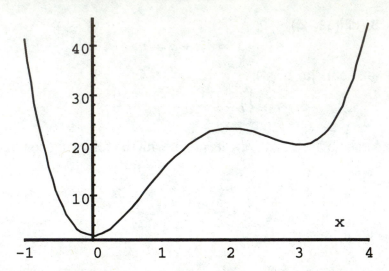

From the graph it would seem that the extrema include relative minima at x = 0 and x = 3, and a relative maximum at x = 2. We can confirm these approximations by finding where the derivative is zero.

• q6 := diff(q5, x);

$$q6 := 8\,x^3 - 40\,x^2 + 48\,x$$

• q7 := solve(q6 = 0, x);

$$q7 := 0, 3, 2$$

The critical points are indeed 0, 2, and 3. We can reinforce the suggestions of the graph by performing the second derivative test at each critical point.

• q8 := diff(q5, x, x);

$$q8 := 24\,x^2 - 80\,x + 48$$

• subs(x = 0, q5);
  subs(x = 0, q8);

$$2$$

$$48$$

The point (0, 2) is therefore a relative minimum.

• subs(x = 2, q5);
  subs(x = 2, q8);

$$\frac{70}{3}$$

-16

The point (2, 70/3) is therefore a relative minimum.

- subs(x = 3, q5);
  subs(x = 3, q8);

20

24

The point (3, 20) is therefore a relative minimum.

To determine where y is decreasing, we need to see where y'(x) < 0. From the graph of y(x) we suspect that y is decreasing when x < 0 and when 2 < x < 3. There are several ways to verify this suspicion. First, we can look at a graph of y'(x) and see where y' < 0.

- plot(q6, x = -1..4);

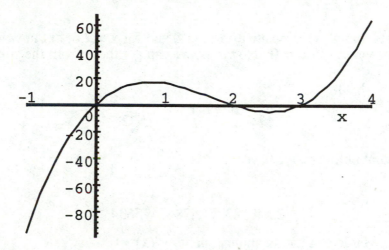

It certainly seems that y'(x) is negative for x < 0, and for 2 < x < 3. We can actually be a bit more certain since we know that y'(0) = y'(2) = y'(3) = 0.

A second way to determine where y'(x) < 0 is by seeing if Maple's **solve** command can deal with the appropriate inequality.

- solve(q6<0, x);

$$\{x<0\}, \{x<3, 2<x\}$$

Maple has returned the same information as the graph did; only the inequalities in {x < 3, 2 < x} have to be interpreted as x < 3, but 2 < x, to obtain the interval 2 < x < 3.

The question on concavity can be answered in exactly the same way. We can either look at a graph of y"(x) to see where it is positive, or we can see if Maple's **solve** command will deal with the inequality y"(x) > 0.

- plot(q8, x = -1..4);

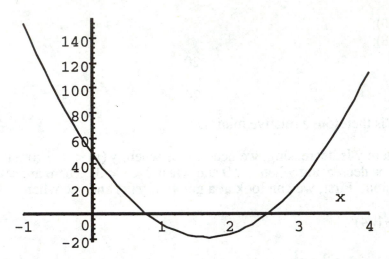

The graph suggests that y"(x) is positive for x < .8 and for x > 2.6, but now we don't have the exact values of where y"(x) = 0. However, we can get these from the **solve** command.

- q9 := solve(q8 = 0, x);

$$q9 := \frac{5}{3} + \frac{1}{3}\sqrt{7}, \frac{5}{3} - \frac{1}{3}\sqrt{7}$$

And as floating-point numbers, we have

- evalf(q9);

$$2.548583771, .7847495634$$

Finally, let's see if Maple can handle the inequality y"(x) > 0.

- solve(q8 > 0, x);

$$\left\{ x < \frac{5}{3} - \frac{1}{3}\sqrt{7} \right\}, \left\{ \frac{5}{3} + \frac{1}{3}\sqrt{7} < x \right\}$$

We have gotten exactly the same solution in two different ways.

Example 4

Find the absolute minimum of f(x) = 1 - 2 tan(x) - 3 cot(x) for x in the closed interval [1.7, 3].

Immediately, begin with a graph.

- q10 := 1 - 2*tan(x) - 3*cot(x);

$$q10 := 1 - 2\tan(x) - 3\cot(x)$$

- plot(q10, x = 1.7..3, 0..13);

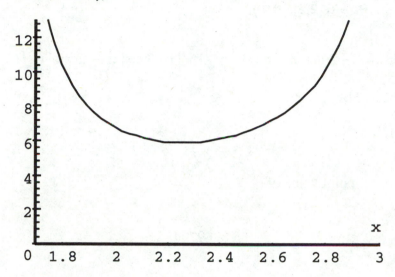

It is clear from the graph that there is a single minimum in the interval [1.7, 3] and that this minimum will be an absolute minimum. The exact location of the minimum will be a zero of the derivative f'(x).

- q11 := diff(q10, x);

$$q11 := 1 - 2\tan(x)^2 + 3\cot(x)^2$$

- q12 := solve(q11 = 0, x);

$$q12 := -\arcsin\left(\frac{1}{5}\sqrt{15}\right), \arcsin\left(\frac{1}{5}\sqrt{15}\right)$$

It appears that we have a first and a fourth quadrant solution when we needed a second quadrant one. We can verify that neither of the solutions in q12 is in the required interval [1.7, 3].

- evalf(q12);

-.8860771237, .8860771237

Rejoice! All the trigonometry learned in the past is still required! The second solution in q12 is a first quadrant solution and is the related angle for the second quadrant angle given by π - arcsin(($\sqrt{15}$)/5).

- evalf(Pi - q12[2]);

$$2.255515530$$

The y-coordinate corresponding to the critical value is obtained by substitution.

• q13 := subs(x = Pi - q12[2], q10);

$$q13 := 1 - 2\tan\left(\pi - \arcsin\left(\frac{1}{5}\sqrt{15}\right)\right) - 3\cot\left(\pi - \arcsin\left(\frac{1}{5}\sqrt{15}\right)\right)$$

This can be compacted by

• simplify(q13);

$$1 + 2\sqrt{3}\sqrt{2}$$

which has the floating-point value of

• evalf(q13);

$$5.898979484$$

Example 5

Find the point on the graph of $f(x) = x^2 \sin(x)$ that is closest to the point (1, 3).

First, let's look at a graph of the function $f(x)$.

• q14 := x^2 * sin(x);

$$q14 := x^2 \sin(x)$$

• plot(q14, x = -3..3);

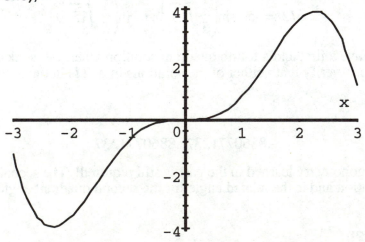

From this graph, the experienced eye might begin to visualize where the minimizing point might be found on f(x). However, the purpose of these exercises is to develop an appropriate intuition that comes only from familiarity and experience.

We need the function representing the distance from the point (1, 3) to an arbitrary point on the graph of f(x). Such an arbitrary point has coordinates (x, y) where y is given by the function f(x). Hence, in *this* Maple session we would represent this arbitrary point on the graph of f(x) as (x, q14).

- dist := sqrt((1 - x)^2 + (3 - q14)^2);

$$dist := \sqrt{(1-x)^2 + \left(3 - x^2 \sin(x)\right)^2}$$

And next, the inevitable plot of the distance function.

- plot(dist, x = -3..3);

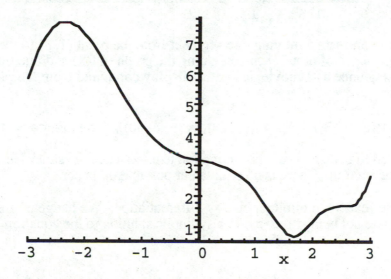

This graph shows that somewhere in the interval [1, 2] there is an x-coordinate for which the distance function is a minimum. Before we compute this minimizing x, let's continue with some graphical explorations. First, let's create a picture showing the graph of f(x) and some representative lines from the point (1, 3) to the graph of f(x). To plot line segments in Maple, it is sufficient to give the **plot** command the coordinates of the endoints of the segment. In fact, let's create a function whose output is the pair of endpoint coordinates for such a line segment.

- S := t-> [[1,3], [t, subs(x = t, q14)]];

$$S := t \rightarrow [[1,3], [t, \text{subs}(x = t, q14)]]$$

The plot we want can now be generated via

- plot({q14, S(1/2), S(1), S(3/2), S(2)}, x = 0..3);

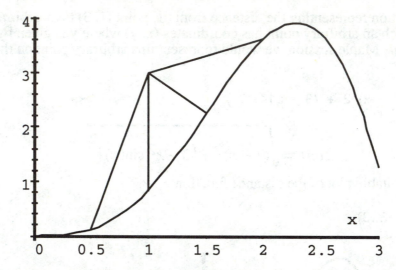

It is even possible to animate a moving line segment from the point (1, 3) to the graph of f(x). All we need is a sequence of plots, each containing the graph of f(x) and one line segment. We then display this sequence as a movie, using the **display** command from the plots package.

- with(plots):
- display([seq(plot({q14, S(.1*k)}, x = 0..3), k = 0..30)], insequence = true);

The reader is invited to code and run this animation to experience the same "ahh" as the authors did upon seeing the motion. There is no equivalent possible on paper.

Finally, we find the "exact" minimum point by differentiation. We hedge on the word "exact" because we anticipate not being able to find a symbolic solution to the equation

- q15 := diff(dist, x) = 0;

$$q15 := \frac{1}{2} \frac{-2 + 2x + 2^{(3 - x^2 \sin(x))}(-2x\sin(x) - x^2\cos(x))}{\sqrt{(1-x)^2 + (3 - x^2\sin(x))^2}} = 0$$

Since we know the minimum is located in the closed interval [1, 2], we can pass this information to **fsolve**, the numeric solver in Maple.

- fsolve(q15, x, x = 1..2);

$$1.671422492$$

Example 6

The numeric solver in the **fsolve** command is essentially a Newton algorithm. It is worthwhile exploring the idea behind Newton's Method, even if it can be invoked automatically by **fsolve**. In fact, there are times when **fsolve** fails and times when Newton's Method fails. Knowing the characteristics of the method will help craft alternative strategies when needed.

Conceptually, Newton's Method arises when the function f(x) in the equation f(x) = 0 is replaced by the line tangent to f(x). Pictorially, this is illustrated by the following diagram, created with the accompanying Maple code. The **textplot** and **display** commands from the plots package are used to label the diagram produced by the **showtangent** command of the student package.

- with(student):
- q16 := showtangent(x^2/2, x = 1,x=0..3/2,0..3/2,scaling=constrained):
  q17 := textplot({[1.4,.5,`(x1,f(x1))`], [1.1,.1,`x1`], [.7,.1,`x2`]}):
  display({q16,q17});

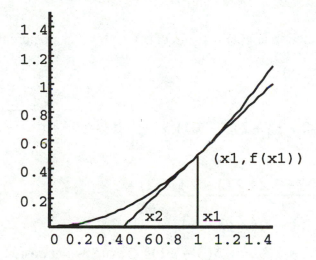

The equation of the tangent line touching at (x1, f(x1)) is y - f(x1) = f'(x1)(x - x1), and this is to be solved for the value of x when y = 0. This value of x is called x2 and is usually a better approximation of the root than x1. Using ff(x1) for f'(x1), we have, in Maple,

- q18 := y - f(x1) = ff(x1)*(x-x1);
$$q18 := y - \mathrm{f}(x1) = \mathrm{ff}(x1)(x - x1)$$

- q19 := subs(y = 0, x = x2, q18);
$$q19 := -\mathrm{f}(x1) = \mathrm{ff}(x1)(x2 - x1)$$

- isolate(q19, x2);

$$x2 = -\frac{f(x1)}{ff(x1)} + x1$$

We used the **isolate** command instead of the **solve** command because **isolate** writes an equation instead of just providing the "answer." The formula for x2 is applied repeatedly, with each new "x2" generally a better approximation of the root than the previous.

Before the availability of symbolic programs like Maple the utility of Newton's Method was limited by the difficulty of calculating and entering the derivative. This no longer being the case, we seek to implement Newton's Method as simply as possible. For our function f(x), let's use the left-hand side of equation q15 from Example 5. It is sufficiently complex that its derivative would be a challenge to obtain and code correctly.

- q20 := lhs(q15);

$$q20 := \frac{1}{2}\frac{-2 + 2\,x + 2\,(3 - x^2\,\sin(x))\,(\text{-}2\,x\,\sin(x) - x^2\,\cos(x))}{\sqrt{(1 - x)^2 + (3 - x^2\,\sin(x))^2}}$$

The Newton iteration function is g(x) = x - f(x)/f'(x), and this can be created in Maple by

- g := x - q20/diff(q20, x);

$$g := x - \frac{1}{2}(-2 + 2\,x + 2\,(3 - \%1)\,\%2) \Big/ \Bigg(\sqrt{(1 - x)^2 + (3 - \%1)^2} \Bigg($$

$$-\frac{1}{4}\frac{(-2 + 2\,x + 2\,(3 - \%1)\,\%2)^2}{((1 - x)^2 + (3 - \%1)^2)^{3/2}}$$

$$+\frac{1}{2}\frac{2 + 2\,\%2^2 + 2\,(3 - \%1)\,(\text{-}2\,\sin(x) - 4\,x\,\cos(x) + \%1)}{\sqrt{(1 - x)^2 + (3 - \%1)^2}} \Bigg)\Bigg)$$

$$\%1 := x^2\,\sin(x)$$

$$\%2 := \text{-}2\,x\,\sin(x) - x^2\,\cos(x)$$

As promised, we have a glimpse of the complexity of g(x). But now, our task is selecting a starting value for the iteration and then devising a simple way of continuing. Suppose we start with x = 1.5, which is in the interval [1, 2].

- xnew := evalf(subs(x = 1.5, g));

$$xnew := 1.789966751$$

The repetitive task of substituting xnew again and again can be handled by a loop.

- for k from 1 to 5 do
      xnew := evalf(subs(x = xnew, g));
      print(k, xnew);
  od:

$$1, 1.611629034$$

$$2, 1.674871445$$

$$3, 1.671415227$$

$$4, 1.671422492$$

$$5, 1.671422492$$

In addition to noting that this iteration produces the same value as **fsolve** did in Example 5, we point out the role of the colon on the loop-ending **od**. The punctuation on the **od** determines whether or not computations inside the loop will echo to the screen. With the colon, there is no output, so a **print** statement is used inside the loop to obtain the desired output.

Example 7

A staple of a calculus course is the notion of a related rate. Probably introduced years ago as an example of the Chain Rule for differentiation, these problems have assumed an identity of their own, largely because students have difficulty with formulating word problems in general. A typical related rate problem would be stated as follows.

A balloon is released at a point 100 feet from a camera that is mounted at ground level. If the balloon goes straight up at the rate of 6 feet per second, (a) how fast is the distance from the camera to the balloon increasing when the balloon is 75 feet high and (b) how fast is the camera turning at the end of 5 seconds?

A labeled diagram is always a good idea. The physical system can be represented as a right triangle with the camera at point A and the balloon rising along the leg CB. We have used L for the hypotenuse AB, x(t) for the time-varying distance CB, and w for the angle of elevation to the balloon.

- q21 := plot({[[0,0],[1,1]], [[1,0],[1,1]], [[0,0],[1,0]]}, axes=none):
  q22 := textplot({[.5,-.1,`100`],[1.2,.6,`x(t)`],[1.2,.4,`x'(t)  =  6`], [.5,.7,`L`],
                    [.2,.1,`w`],[-.05,.05,`A`],[1.05,.05,`C`], [1,1.1,`B`]}):
  display({q21,q22});

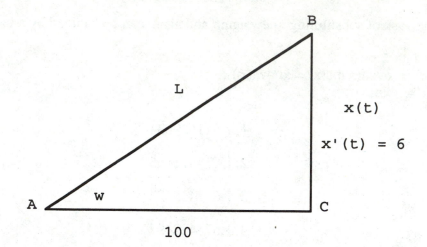

In part (a) we are to find how the distance L varies in time. Clearly, we need to formulate an expression for L.

- L := sqrt(100^2 + x(t)^2);

$$L := \sqrt{10000 + x(t)^2}$$

The question about how fast is the distance L changing means we want the derivative of L with respect to time t. This differentiation is now trivial since we have built into L the essential recognition that x = x(t).

- Ldot := diff(L, t);

$$Ldot := \frac{x(t)\left(\frac{\partial}{\partial t}x(t)\right)}{\sqrt{10000 + x(t)^2}}$$

In Ldot we have x(t) = 75, and x'(t) = 6. All that's left is a substitution in which the set braces are essential for the replacements to be made in the correct order.

- q23 := subs({x(t) = 75, diff(x(t),t) = 6}, Ldot);

$$q23 := \frac{18}{625}\sqrt{15625}$$

- simplify(q23);

$$\frac{18}{5}$$

Part (b) requires that we express the angle w in terms of the varying quantity x(t). Since tan(w) = x(t)/100, we have that w is given by

- w := arctan(x(t)/100);

$$w := \arctan\left(\frac{1}{100} x(t)\right)$$

The rate of change of w is just the derivative dw/dt which is now simple since we anticipated the t-dependence through x(t).

- wdot := diff(w, t);

$$wdot := \frac{1}{100} \frac{\frac{\partial}{\partial t} x(t)}{1 + \frac{1}{10000} x(t)^2}$$

The added complication in this part is the need to determine a value for x(t) at time t = 5 seconds. However, since x(t) is varying at a constant rate of 6 feet per second, at the end of 5 seconds x(t) must be 30 feet. Thus,

- q24 := subs({x(t) = 30, diff(x(t),t) = 6}, wdot);

$$q24 := \frac{6}{109}$$

The number 6/109 is an angular velocity expressed in radians per second. A typical real-world example of the need for such a quantity would be in filming the ascent of a rocket at Cape Kennedy. The rate of rotation of the camera would need to be great enough to match the rate of climb of the rocket.

Exercises

1. For $f(x) = \dfrac{1}{x^2 + 1}$, find

    (a) The average rate of change of f(x) with respect to x as x changes from 1 to 3.

    (b) The instantaneous rate of change of f(x) with respect to x at x = 1.

2. Find the derivative of

    (a) $\dfrac{1}{\sqrt{x + 1}}$ **(b)** $(x^3 + 8)^4$ **(c)** $\cos(x^2 + 3x)$.

3. Find the real roots of $4x^3 - 10x^2 - 20x + 10 = 0$.

4. Find a point on the graph of $y = 15x^3 - 24x^2 - 101x + 150$ where the tangent line is parallel to x - 3y = 5.

5. Find a point on the graph of $y = \dfrac{3}{2}x^4 - 4x^3 - 12x^2 + 24x - 1$ where the tangent line is perpendicular to x - 2y = 1.

6. Find the smallest real root of $x^3 \sin x + x = 0$ in $[\pi, +\infty)$.

7. Where, on the graph of $f(x) = \left(\dfrac{3x + 1}{x^2}\right)^3$, is the tangent line parallel to the x-axis?

8. Suppose that you are making a video of a bicycle race from a position that is 24 feet from the edge of a straight road. If a cyclist passes by at 27 mph, how fast must you turn your camera to track this racer

    (a) when the bike is directly in front of you?

    (b) when the bike is 1 second past?

    (c) when the bike is 5 feet past?

9. A balloon is 200 feet off the ground and rising at the rate of 15 feet per second. A bicycle passes beneath it traveling in a straight line at 66 feet per second. How fast is the distance between them changing

    (a) 1 second later?   **(b)** 10 seconds later?

10. Corn pouring from a tube at the rate of 10 cubic feet per minute forms a conical pile having altitude 1.3 times the radius of the base. How fast is the altitude of the pile changing when the pile is 12 feet high?

11. Find an equation of the line tangent to the graph of $y = x^5 - 5x^4 - 3x^3 + 30x^2 + 3$ at the rightmost inflection point of the curve.

12. Determine how many horizontal tangents exist for each function:

    (a) $y = x^3 + 3x^2 + 2x + 43$ (c) $y = 2x^3 + x^2 + x - 5$.

    (b) $y = 3x^3 + 6x^2 + 4x$

13. Define $f(x) = 50x^4 - 202x^3 - 397x^2 + 2393x + 480$.

    (a) Locate the critical points.

    (b) Classify each critical point.

    (c) Plot the graph on an appropriate domain/range.

14. Find the maximum and minimum values of $y = \sin x + 2\sin 2x$.

15. Find the extrema of $f(x)$ on the interval $[-4, 5]$ for

    (a) $f(x) = \cos x - \sqrt{4 + x}$ (b) $f(x) = 3\cos x - \sqrt{4 + x}$.

16. Find (if possible) the maximum and minimum values of $y = \dfrac{24}{\sin x} + \dfrac{5}{\cos x}$ on

    (a) $(0, \frac{\pi}{2})$ (b) $[\frac{\pi}{4}, \frac{\pi}{3}]$.

17. A weight attached to a spring moves according to $y = 2\sin 3t + \cos 2t$, where y measures displacement from the equilibrium point which is at the origin.

    (a) At what times is the velocity $= 0$?

    (b) Find the maximum displacement from the origin.

18. Find values of a, b, c and d such that $f(x) = ax^3 + bx^2 + cx + d$ has a relative minimum at $(-2, -1)$ and a relative maximum at $(3, 4)$.

19. Let $f(x) = 2x^3 - 3x^2 - 12x + 12$.

(a) Where is f(x) decreasing?

(b) Where are the relative extrema of f(x)?

(c) Where is f(x) concave up?

(d) Confirm your answers visually with a plot.

20. Let $f(x) = 2 x^3 - 4 x^2 - 12 x + 12$.

(a) Where is f(x) decreasing?

(b) Where are the relative extrema of f(x)?

(c) Where is f(x) concave up?

(d) Confirm your answers visually with a plot.

21. Let $f(x) = 3 x^4 - 8 x^3 - 30 x^2 + 72 x + 24$.

(a) Where is f(x) decreasing?

(b) Where are the relative extrema of f(x)?

(c) Where is f(x) concave up?

(d) Confirm your answers visually with a plot.

22. Let $f(x) = 3 x^4 + 8 x^3 - 30 x^2 + 72 x + 24$.

(a) Where is f(x) decreasing?

(b) Where are the relative extrema of f(x)?

(c) Where is f(x) concave up?

(d) Confirm your answers visually with a plot.

23. Find the vertical and horizontal asymptotes of $y = \dfrac{x^{1/3}}{x^{2/3} - 5}$. Locate any relative extrema and plot the graph using a domain/range that shows the features of the function clearly.

24. The comet Zagon follows a planar trajectory whose equation is $y = x^2 \tan x$. In the same coordinate system, the planet Zyfor is located at the point having coordinates (2, 1). Plot the graph of the trajectory of Zagon, and three line segments from the location of Zyfor to

different points on the trajectory. What is the least distance between Zyfor and the orbit of Zagon?

25. Refer to Exercise 24. Create an animation which shows Zagon as it follows its orbit in the vicinity of Zyfor. Can you leave a marker at the point on the orbit which is nearest Zyfor?

26. Plot the graph of $f(x) = x^2 \tan x$ and the point $P = (2.1, 0.8)$ in the same window. Find the point on the graph of $f(x)$ which is closest to P.

27. Plot the graph of $f(x) = x^3$ and the point $P = (1, 0)$ in the same window. Find the point on the graph of $f(x)$ which is closest to P.

28. Plot the graph of $f(x) = x^3 \sin x$ and the point P in the same window. Find the point on the graph of $f(x)$ which is closest to P

   (a) $P = (\frac{\pi}{2}, 0.1)$ (b) $P = (2.34, 2)$.

29. Plot the graph of $f(x) = x^4 \sin x$ and the point $P = (1, -0.5)$ in the same window. Find the point on the graph of $f(x)$ which is closest to P.

30. Make a 13-frame movie of the plot of $f = \dfrac{x^3}{3} - 2x^2 + 3x + 1$ for $0 \le x \le 4$.

31. Modify your movie in Exercise 30 to include the plot of f'. What happens to the value of f' when f reaches the peak at $x = 1$?

32. Find the dimensions of the rectangle of maximum area that can be inscribed in the ellipse $\dfrac{x^2}{25} + \dfrac{y^2}{9} = 1$.

33. A ladder is to lean over a 6 foot high fence and reach a wall that is 1 foot behind the fence. What is the length of the shortest ladder that will reach the wall?

34. Find the first quadrant point P on the parabola $y = 2 - x^2$, for which the triangle enclosed by the tangent line at P, and the coordinate axes, has minimal area.

35. Find the first quadrant point P on the curve $y = 4 - x^3$, for which the triangle enclosed by the tangent line at P, and the coordinate axes, has minimal area.

36. (a) Use Newton's Method to approximate the smallest root of $x^2 \sin x + 1 = 0$ in the interval $[\pi, +\infty)$. Stop iterating when the error is less than 0.01.

   (b) Check your result using **fsolve** with options.

37. (a) Use Newton's Method to approximate the smallest positive root of $x^2 + \tan x = 0$ in the interval $[\pi, +\infty)$. Stop iterating when the error is less than 0.005.

    (b) Check your result using **fsolve**.

38. The position of a moving particle is given by $s = \dfrac{t+1}{t^2+6}$. At what time will the particle first reverse its direction? At what time will it next reverse its direction?

39. Pick a point P on the parabola $y = x^2$. Let $Q(\neq P)$ be the intersection of the parabola with the normal line to the parabola at P. For what choice of P is the distance between P and Q minimum?

Projects

1. Let $f(x) = 13x - 5x^2 + 15$.

    (a) Find the critical point $(x_0, f(x_0))$ of $f(x)$.

    (b) Compute $f(x_0) - f(x_0 + h)$ for $h = +0.1, +0.01$ and $+0.001$.

    (c) Compute $f(x_0) - f(x_0 + h)$ for $h = -0.1, -0.01$ and -0.001.

    (d) Your results in parts (b) and (c) should agree in sign. Does this suggest that $f(x_0)$ is a relative maximum or a relative minimum?

    (e) Since each value in parts (b) and (c) has the same sign, does this prove that $f(x_0)$ is a relative maximum (or a relative minimum)? What could happen?

    (f) Plot $f(x)$ on $[0, 3]$ to check your answer to part (d).

2. This problem is similar to Problem 1, but deals with the function e^x, that is given in Maple by exp(x).

    Define $g(x) = e^x + \dfrac{1}{5}\cos(20x)$.

    (a) Find the critical point x_0 in the interval $[0.95..1]$.

    (b) Compute $g(x_0) - g(x_0 + h)$ for several small values of h. (Both + and -.)

(c) Use the information from part (b) to predict whether $g(x_0)$ is a relative maximum or a relative minimum.

(d) Plot $g(x)$ over an interval that includes x_0. Does the plot support your determination of a relative maximum or a relative minimum?

3.   Define $f(x) = 12 x^5 - 120 x^4 + 400 x^3 - 480 x^2 + 5$.

(a) Find the three critical points of $f(x)$.

(b) For each critical point x_i, compute $f(x_i) - f(x_i + h)$ for several small values of h, including positive and negative values.

(c) Use the information from part (b) to classify $f(x_i)$ as a relative maximum, a relative minimum, or neither.

(d) Plot $f(x)$ over an interval that includes all three critical points. Does your plot support your conclusion from part (c)?

(e) Factor the derivative of $f(x)$. Why does the square factor cause the function to "lose" the relative extrema behavior at one of the critical points?

4.   A man bought for his son a new type of bicycle speedometer, which included a 1-meter trailing arm connected to a 200 mm diameter sensor wheel rolling on the ground. Not being an engineer, the man mistakenly attached the arm to a pedal rather than to the bicycle frame. The bicycle wheels have a diameter of 660 mm, and each pedal crank has a length of 160 mm with the crank axis being 270 mm above the ground. The pedal sprocket has 48 teeth and the wheel sprocket has 16. What is the maximum linear speed of the sensor wheel, to the nearest mm/sec., at a pedal rate of 100 rpm?
                                        -R. Wilson Rowland (The Bent)

5.   To approximate the reciprocal of a positive number C, define $f(x) = \frac{1}{x} - C$ and apply Newton's Method. (Note that no division is required.) Use this method to approximate to three places

(a) $\frac{1}{2}$ (b) $\frac{1}{7}$ (c) $\frac{1}{99}$.

6.   Find three functions and a starting x_0 for each which behave in the following ways when Newton's Method is used to approximate a root of the resulting equation.

(a) The iterates cycle, that is, $x_0 = x_2 = x_4 = ...$ and $x_1 = x_3 = x_5 = ...$.

(b) The sequence of iterates diverges, for example, $x_0 < x_1 < x_2 ... \rightarrow +\infty$.

(c) Depending on the starting value x_0, the sequence of iterates converges or diverges. Find the exact value of a such that x_0 to one side of a causes convergence and x_0 to the other side causes divergence. Plot the function and the first three or four iterates in each case. Two functions are given which will help you with this problem.

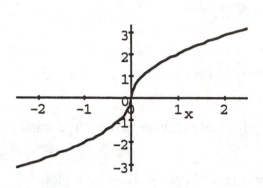

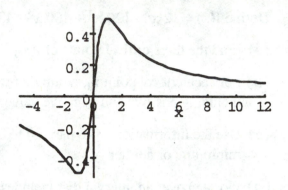

$$f(x) = \begin{cases} 2\sqrt{x}, & x > 0 \\ -2\sqrt{-x}, & x \le 0 \end{cases} \qquad\qquad g(x) = \frac{x}{(x^2 + 1)}$$

A piecewise defined function such as f(x) can be entered into Maple via the syntax
f := x-> if x > 0 then 2*sqrt(x) else -2*sqrt(-x) fi:

7.   Between towns A and B lies a wedge-shaped marsh, as shown in the figure. AC and BD are the perpendicular distances to the edge of the marsh. AC = 4 km, OC = 5 km, BD = 2 km, OD = 10 km, and angle COD = 30 degrees. The cost of a road is $400,000 per kilometer on dry land and $800,000 per kilometer through the marsh. What is the minimum cost of a road connecting the two towns?

-J.W. Langhaar (TheBent)

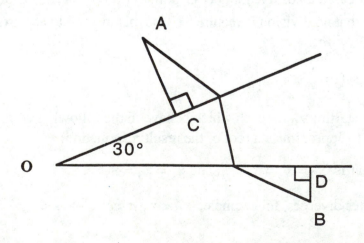

8.   Define f: $= x^3 - 2x^2 + x - 3$. Plot f for $1 \le x \le 4$. Apply 3 iterations of Newton's method to solve the equation f(x) = 0, starting at $x_0 = 3.5$. Animate this process to show how the tangent line is intersected with the x-axis to yield successive iterates.

9.   Asymptotes: Horizontal, Vertical, and Otherwise

(a) Define $f(x) = \dfrac{x^3 + 5x + 8}{x + 3}$ and plot f(x) for $-10 \le x \le 10$, $-10 \le y \le 10$. Observe that there is a vertical asymptote but no horizontal asymptote. Now write f(x) as a quotient $q = x + 2$ plus a remainder $r = \dfrac{2}{x + 3}$. Display f(x) and q in the same window for $-10 \le x \le 10$, $-10 \le y \le 10$. f(x) is asymptotic to q, and q is said to be an oblique asymptote. Compute the vertical distance between f(x) and q. What happens to this distance when x is large?

(b) Define $f(x) = \dfrac{2x^4 + 7x + 4}{x + 3}$. Find any relative extrema, intercepts, and asymptotes. Plot f(x) and its asymptotes in the same window.

(c) Define $f(x) = \dfrac{x^3 - x^2 + 6x - 2}{2x - 2}$. Find any relative extrema, intercepts, and asymptotes. Plot f(x) and its asymptotes in the same window Is the asymptote determined by the quotient a straight line? How can you recognize the asymptotes associated with a rational function f(x)?

(d) Devise a function f(x) that is asymptotic to $q = x + 1$ and has a vertical asymptote at $x = 0$. Display f(x) and q.

(e) Devise a function f(x) that is asymptotic to $q = x^2 - 2x$ and has a vertical asymptote at $x = -2$. Display f(x) and q.

Chapter 5

Integration

New Maple Commands for Chapter 5

denom(fraction);	returns denominator of fraction
int(f(x),x);	indefinite integral (antiderivative of f(x))
int(f(x),x=a..b);	definite integral of f(x) between a and b
Int(f(x),x);	inert integral, not evaluated
sec(x);	secant of x in radians
sum(k,k=1..3);	returns $1 + 2 + 3$
Sum(k,k=1.3);	returns sigma notation for the sum $1 + 2 + 3$
with(student);	loads student package
integrand(Int(f,x));	returns integrand of $\int f\, dx$
leftbox(f(x),x=a..b,n);	graphs f(x) on [a, b], drawing n rectangles under f(x) to approximate area
leftsum(f(x),x=a..b,n);	exact sum of areas of rectangles from leftbox

---

Introduction

If area under a curve is the starting point in a discussion of the Integral Calculus, then it is likely that area by a summation of rectangles is the first activity in such a study. As soon as the Fundamental Theorem of Calculus is established the area problem becomes identified as an antidifferentiation question. At this point, the integral symbol introduced as the signpost for the summation process becomes identified with the antiderivative.

Since Maple has an **int** command that delivers antiderivatives, but not an antiderivative command per se, we will not distinguish between antidifferentiation and the **int** operator. Hence, all uses of the **int** command will be synonymous with antidifferentiation. In the course of compositing the integrals in these examples, we will make use of Maple's inert integration operator **Int** which creates unevaluated integrals that can be evaluated by the **value** command. Finally, we'll make use of the student package which contains a number of functions useful in learning about integration.

Examples

Example 1

- with(student):
- q1 := Int((x^2 - 2), x);

$$q1 := \int x^2 - 2 \, dx$$

We will comment only this once that Maple really should use parentheses around the integrand.

- q2 := value(q1);

$$q2 := \frac{1}{3}x^3 - 2x$$

- diff(q2, x);

$$x^2 - 2$$

The integration (i.e., the antidifferentiation) could have been performed via the **int** command without the intermediate step of writing the inert integral to the screen. Moreover, the differentiation of the antiderivative in q2 restores the integrand in q1.

Example 2

- q3 := Int((x^3 + 4)^6 * x^2, x);

$$q3 := \int (x^3 + 4)^6 x^2 \, dx$$

- q4 := value(q3);

$$q4 := \frac{1}{21}x^{21} + \frac{4}{3}x^{18} + 16x^{15} + \frac{320}{3}x^{12} + \frac{1280}{3}x^9 + 1024x^6 + \frac{4096}{3}x^3$$

- q5 := diff(q4, x);

$$q5 := x^{20} + 24x^{17} + 240x^{14} + 1280x^{11} + 3840x^8 + 6144x^5 + 4096x^2$$

To compare the result to the original integrand requires that we either expand the integrand in q3 or factor the differentiation in q5.

- expand(integrand(q3));

$$x^{20} + 24x^{17} + 240x^{14} + 1280x^{11} + 3840x^8 + 6144x^5 + 4096x^2$$

- factor(q5);

$$(x^3 + 4)^6 x^2$$

Both approaches verify that integration and differentiation are inverse operators.

Example 3

- q6 := Int((2*x - 3)^5, x);

$$q6 := \int (2\,x - 3)^5\,dx$$

• q7 := value(q6);

$$q7 := \frac{1}{12}(2\,x - 3)^6$$

• diff(q7, x);

$$(2\,x - 3)^5$$

Again, the derivative of the antiderivative is the integrand.

Example 4

• q8 := Int(sin(x)^2 * cos(x)^2, x);

$$q8 := \int \sin(x)^2 \cos(x)^2\,dx$$

• q9 := value(q8);

$$q9 := -\frac{1}{4}\sin(x)\cos(x)^3 + \frac{1}{8}\cos(x)\sin(x) + \frac{1}{8}x$$

• diff(q8, x);

$$\sin(x)^2 \cos(x)^2$$

Once again, the derivative of the antiderivative is the integrand.

Example 5

The **int** command is also used for definite integration.

• q10 := Int(sin(x) * cos(x), x = 0..Pi/2);

$$q10 := \int_{0}^{\frac{1}{2}\pi} \sin(x)\cos(x)\,dx$$

• value(q10);

$$\frac{1}{2}$$

It is a useful learning experience to obtain this result via the appropriate indefinite integral (i.e., via an antiderivative) and evaluation at the limits. Thus

• f := integrand(q10);

$$f := \sin(x)\cos(x)$$

• q11 := int(f, x);

$$q11 := \frac{1}{2}\sin(x)^2$$

The expression in q11 is the appropriate antiderivative. Evaluation at the limits is done as follows.

• q12 := subs(x = Pi/2, q11) - subs(x = 0, q11);

$$q12 := \frac{1}{2}\sin\left(\frac{1}{2}\pi\right)^2 - \frac{1}{2}\sin(0)^2$$

• simplify(q12);

$$\frac{1}{2}$$

It is no surprise that both computations resulted in the same value of 1/2.

Example 6

One method of integration that is a staple for a traditional integral calculus course deals with rational functions; hence, an example of the behavior of a rational function. Recall, however, that every zero of the denominator is a potential site for a vertical asymptote. So, after entering the rational function, it will be useful to determine where the candidates for asymptotes might be.

• f := (100*x^3 + 100*x^2 - 25*x - 26)/
          (100*x^4 - 445*x^3 + 44*x^2 + 1780*x - 1800);

$$f := \frac{100\,x^3 + 100\,x^2 - 25\,x - 26}{100\,x^4 - 445\,x^3 + 44\,x^2 + 1780\,x - 1800}$$

• fsolve(denom(f), x);

          -2.003451417, 1.662402617, 1.817585972, 2.973462829

The potential for having two vertical asymptotes so close together (1.66 and 1.8) means a naive reliance on the **plot** command will leave the reader shorthanded. In fact, note the very unsatisfactory result obtained by

• plot(f, x = -3..4, -5..5);

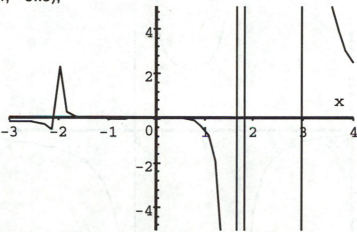

Maple struggles with the asymptote near x = -2, and the closely spaced asymptotes between x = 1.6 and 1.8 show no points of the function between. Similarly, between the asymptotes near x = 1.8 and x = 3, we also see no points on the graph. Incidentally, the second range in the plot command limits the height of the points being plotted, a strategy often needed in the presence of vertical asymptotes. However, here, even this ploy is not enough to generate an acceptable plot.

If we try telling Maple to use more points via the *numpoints* option, we get a slightly improved plot, but not one improved enough to be satisfactory.

• plot(f, x = -3..4, -5..5, numpoints = 500);

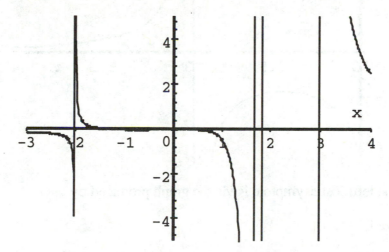

The best way of examining this function graphically is to plot local portions. For example, to see what happens between the two closely spaced asymptotes, try

• plot(f, x = 1.5 .. 2, -500..500);

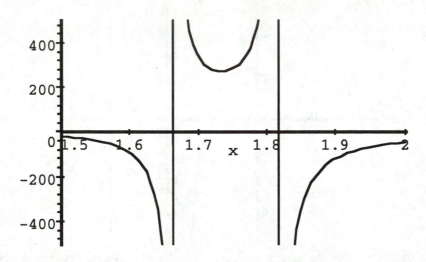

Thus, the function f(x) is a difficult one to plot because the scales vary greatly in different parts of its domain. This plot illuminates the middle two vertical asymptote. Confirmation of the rightmost asymptote is via the graph created by

• plot(f, x = 2..4, -50..50);

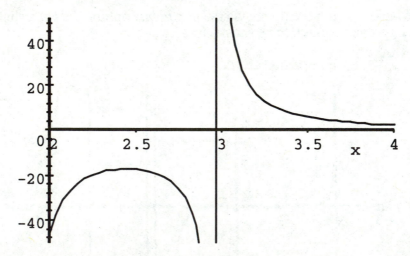

Confirmation of the leftmost asymptote is via the graph produced by

• plot(f, x = -3..-1, -1..1);

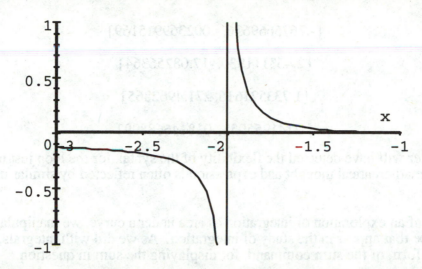

Horizontal asymptotes are obvious by inspection since the degree of the numerator is less than the degree of the denominator. The following limits confirm that the x-axis (y = 0) is a horizontal asymptote.

• limit(f, x = infinity);
  limit(f, x = -infinity);

$$0$$

$$0$$

Intercepts are the zeros of f, which are therefore the same as the zeros of the numerator.

• fsolve(numer(f), x);

$$-.9861593668, -.5204346405, .5065940073$$

Finally, the critical points include the zeros of the derivative of f(x), which will be the same as the zeros of the numerator of the derivative.

• q13 := numer(diff(f, x));

$$q13 := -10000\,x^6 - 20000\,x^5 + 56400\,x^4 + 344150\,x^3 - 395610\,x^2 - 357712\,x + 91280$$

• q14 := fsolve(q13, x);

$$q14 := -.7675669636, .2141955058, 1.733521616, 2.453111830$$

• for s in {q14} do print(subs(x = s, [x, f])); od;

$$[-.7675669636, -.002369915169]$$

$$[2.453111830, -17.08256364]$$

$$[1.733521616, 271.4962865]$$

$$[.2141955058, .01814668899]$$

The astute reader will have detected the flexibility of the syntax for the loop just used. Maple's faithfulness to mathematical thought and expression is often reflected by similar useful constructions.

In anticipation of an exploration of integration as area under a curve, we manipulate several sums of the type that appear in the study of integration. As we did with integrals, we use **Sum**, the inert form of the **sum** command, for displaying the sum in question.

Example 7

a. Evaluate

• q15 := Sum(k/3, k = 1..10);

$$q15 := \sum_{k=1}^{10} \left(\frac{1}{3} k \right)$$

• value(q15);

$$\frac{55}{3}$$

b. Find 1/9 - 1/16 + 1/25 - ... + 1/169.

• q16 := sum((-1)^(k+1) / k^2, k = 3..13);

$$q16 := \frac{9765337663}{129859329600}$$

• evalf(q16);

$$.07519935374$$

Example 8

For the function $f(x) = 10 - x^2$ use the **leftbox** and **leftsum** commands of the student package to approximate the area bounded by $f(x)$, the x-axis, and the line $x = 1$. Use eight subintervals.

- f := 10 - x^2;

$$f := 10 - x^2$$

- leftbox(f, x = 1..sqrt(10), 8);

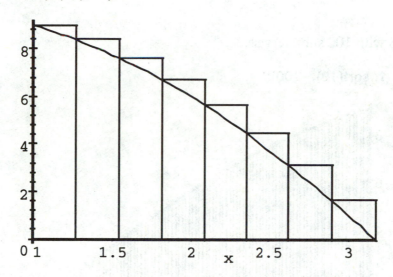

- q17 := leftsum(f, x = 1..sqrt(10), 8);

$$q17 := \left(\frac{1}{8}\sqrt{10} - \frac{1}{8}\right)\left(\sum_{i=0}^{7}\left(10 - \left(1 + i\left(\frac{1}{8}\sqrt{10} - \frac{1}{8}\right)\right)^2\right)\right)$$

- q18 := value(q17);

$$q18 := \left(\frac{1}{8}\sqrt{10} - \frac{1}{8}\right)\left(79 - \left(\frac{7}{8} + \frac{1}{8}\sqrt{10}\right)^2 - \left(\frac{3}{4} + \frac{1}{4}\sqrt{10}\right)^2 - \left(\frac{5}{8} + \frac{3}{8}\sqrt{10}\right)^2\right.$$
$$\left. - \left(\frac{1}{2} + \frac{1}{2}\sqrt{10}\right)^2 - \left(\frac{3}{8} + \frac{5}{8}\sqrt{10}\right)^2 - \left(\frac{1}{4} + \frac{3}{4}\sqrt{10}\right)^2 - \left(\frac{1}{8} + \frac{7}{8}\sqrt{10}\right)^2\right)$$

- expand(q18);

$$\frac{921}{128}\sqrt{10} - \frac{1299}{128}$$

- evalf(q18);

$$12.60513848$$

Example 9

Repeat Example 8 with 100 subintervals.

- leftbox(f, x = 1..sqrt(10), 100);

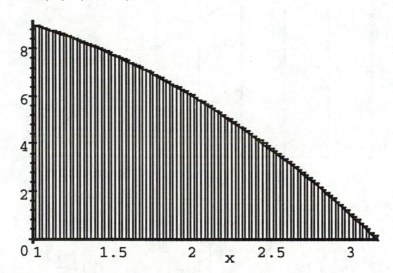

- q19 := leftsum(f, x = 1..sqrt(10), 100);

$$q19 := \left(\frac{1}{100}\sqrt{10} - \frac{1}{100}\right)\left(\sum_{i=0}^{99}\left(10 - \left(1 + i\left(\frac{1}{100}\sqrt{10} - \frac{1}{100}\right)\right)^2\right)\right)$$

- expand(value(q19));

$$\frac{134229}{20000}\sqrt{10} - \frac{194223}{20000}$$

- evalf(value(q19));

$$11.51231840$$

Example 10

Repeat Example 8 with an arbitrary number of subintervals, n. Then, take the limit as n becomes infinite.

- q20 := leftsum(f, x = 1..sqrt(10), n);

$$q20 := \frac{(\sqrt{10} - 1)\left(\displaystyle\sum_{i=0}^{n-1}\left(10 - \left(1 + \frac{i(\sqrt{10} - 1)}{n}\right)^2\right)\right)}{n}$$

- q21 := value(q20);

$$q21 := \frac{(\sqrt{10} - 1)\left(\dfrac{19}{3}n - \dfrac{1}{3}\sqrt{10}\,n + \dfrac{9}{2} - \dfrac{11}{6}\dfrac{1}{n} + \dfrac{1}{3}\dfrac{\sqrt{10}}{n}\right)}{n}$$

- q22 := limit(q21, n = infinity);

$$q22 := -\frac{1}{3}(\sqrt{10} - 1)(-19 + \sqrt{10})$$

- expand(q22);

$$\frac{20}{3}\sqrt{10} - \frac{29}{3}$$

- evalf(q22);

$$11.41518440$$

Example 11

Repeat Example 8 with eight subintervals, this time using the **sum** command to construct the approximating sum yourself. Begin by determining d, the width of a single rectangle.

- d := (sqrt(10) - 1)/8;

$$d := \frac{1}{8}\sqrt{10} - \frac{1}{8}$$

Next, write an expression for x_k, the kth partition point on the x-axis.

- xk := 1 + k*d;

$$xk := 1 + k\left(\frac{1}{8}\sqrt{10} - \frac{1}{8}\right)$$

Since we are still evaluating, at the left edge of the subintervals, the heights of the approximating rectangles, we get for the approximating sum

- q23 := Sum(subs(x = xk, f)*d, k = 0..7);

$$q23 := \sum_{k=0}^{7}\left(10 - \left(1 + k\left(\frac{1}{8}\sqrt{10} - \frac{1}{8}\right)\right)^2\right)\left(\frac{1}{8}\sqrt{10} - \frac{1}{8}\right)$$

- q24 := value(q23);

$$q24 := \frac{9}{8}\sqrt{10} - \frac{9}{8} + \left(10 - \left(\frac{7}{8} + \frac{1}{8}\sqrt{10}\right)^2\right)\%1 + \left(10 - \left(\frac{3}{4} + \frac{1}{4}\sqrt{10}\right)^2\right)\%1$$

$$+ \left(10 - \left(\frac{5}{8} + \frac{3}{8}\sqrt{10}\right)^2\right)\%1 + \left(10 - \left(\frac{1}{2} + \frac{1}{2}\sqrt{10}\right)^2\right)\%1$$

$$+ \left(10 - \left(\frac{3}{8} + \frac{5}{8}\sqrt{10}\right)^2\right)\%1 + \left(10 - \left(\frac{1}{4} + \frac{3}{4}\sqrt{10}\right)^2\right)\%1$$

$$+ \left(10 - \left(\frac{1}{8} + \frac{7}{8}\sqrt{10}\right)^2\right)\%1$$

$$\%1 := \frac{1}{8}\sqrt{10} - \frac{1}{8}$$

- expand(q24);

$$\frac{921}{128}\sqrt{10} - \frac{1299}{128}$$

- evalf(q24);

$$12.60513848$$

Example 12

Compute the area of the region in Example 8, this time exactly, via the Fundamental Theorem of Calculus.

• q25 := Int(f, x = 1..sqrt(10));

$$q25 := \int_{1}^{\sqrt{10}} 10 - x^2 \, dx$$

• q26 := value(q25);

$$q26 := \frac{20}{3}\sqrt{10} - \frac{29}{3}$$

• evalf(q26);

$$11.41518440$$

The answers in Examples 10 and 12 agree, both as floating-point values and as exact (symbolic) values.

Example 13

Find the area bounded by f(x) = tan(x/3) sec(x/3) + 4 sin(2 x) and the x-axis, between x = 0, and x = π.

A graph is essential, for it will reveal that some of the area is above the x-axis (positive) and some is below the x-axis (negative).

• f := tan(x/3)*sec(x/3) + 4*sin(2*x);

$$f := \tan\left(\frac{1}{3}x\right)\sec\left(\frac{1}{3}x\right) + 4\sin(2\,x)$$

• plot(f, x = 0..Pi);

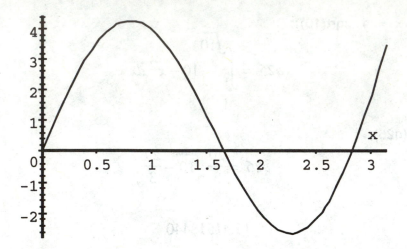

We must determine the zeros of f(x) because the integration that will give us the desired area must be broken at these zeros.

- q27 := fsolve(f, x, x = 1..2);

$$q27 := 1.662255839$$

- q28 := fsolve(f, x, x = 2..3);

$$q28 := 2.828466588$$

- q29 := Int(f, x = 0..q27) - Int(f, x = q27..q28) + Int(f, x = q28..Pi);

$$q29 := \int_{0}^{1.662255839} \%1 \, dx - \int_{1.662255839}^{2.828466588} \%1 \, dx + \int_{2.828466588}^{\pi} \%1 \, dx$$

$$\%1 := \tan\left(\frac{1}{3}x\right) \sec\left(\frac{1}{3}x\right) + 4\sin(2x)$$

- value(q29);

$$7.017187624$$

Example 14

Find the average value of f(x) = x/√(x² + 10) on the interval [5, 8]. Then draw a representative picture that demonstrates the meaning of "average value."

- f := x/sqrt(x^2 + 10);

$$f := \frac{x}{\sqrt{x^2 + 10}}$$

• q30 := Int(f, x = 5..8)/(8 - 5);

$$q30 := \frac{1}{3} \int_5^8 \frac{x}{\sqrt{x^2 + 10}} dx$$

• q31 := value(q30);

$$q31 := \frac{1}{3}\sqrt{74} - \frac{1}{3}\sqrt{35}$$

• evalf(q31);

$$.895415161$$

• plot({f, q31, [[5,q30], [5,0]], [[8,q30], [8,0]]},   x = 5..8);

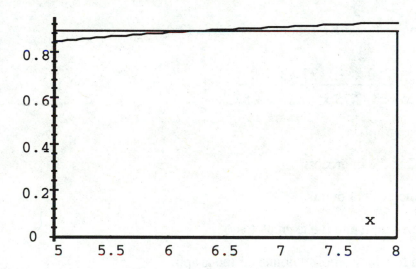

Exercises

In exercises 1 - 10, use Maple to find the antiderivative; then check your answer by differentiating. Can you do some of these easily with paper and pencil?

1. $\int x^3 \, dx$

6   $\int (\frac{1}{\sqrt{x}} + \exp(x)) \, dx$

2. $\int (x^5 - 6x + 12) \, dx$

7. $\int \sec(3x) \tan(3x) \, dx$

3. $\int \sin(3x) \, dx$

8. $\int \tan^2 x \, dx$

4. $\int \tan x \, dx$

9. $\int x\sqrt{x^2 + 4} \, dx$

5. $\int \sin^3 x \cos x \, dx$

10. $\int (3x + 4)^{12} \, dx$

11. Find the smallest positive root of $6x + 25 \sin(2x) = \frac{107}{4}$.

12. Let $f(x) = \dfrac{250 x^2 - 1175 x + 430}{250 x^3 - 575 x^2 - 415 x + 572}$.

   (a) Plot the graph.

   (b) Calculate the intercepts.

   (c) Find the critical points.

   (d) Plot the graph and the asymptotes.

   (e) Find a few well-chosen points on the graph.

13. Let $f(x) = \dfrac{2 x^4 + 2.601 x^3 - 3.399 x^2 - 1.202 x}{x^4 - 2.742 x^3 - 1.349 x^2 + 6.299 x - 1.267}$.

   (a) Plot the graph.

   (b) Calculate the intercepts.

   (c) Find the critical points.

(d) Plot the graph and the asymptotes,

(e) Find a few well-chosen points on the graph.

14. Use the command **sum** to find the value of

(a) $\displaystyle\sum_{k=4}^{10} k^2$

(d) $\displaystyle\sum_{k=1}^{9} (\frac{1}{k} - \frac{1}{k+1})$

(b) $\displaystyle\sum_{k=1}^{15} (2k-3)$

(e) $\displaystyle\sum_{k=1}^{n} (\frac{1}{k} - \frac{1}{k+1})$

(c) $\displaystyle\sum_{k=4}^{8} 2^k$

(f) $1 - \frac{1}{2} + \frac{1}{3} - \frac{1}{4} + \dots - \frac{1}{10}$

15. Use a **for** loop to find the value of

(a) $\displaystyle\sum_{k=1}^{10} k^2$

(c) $\displaystyle\sum_{k=4}^{8} 2^k$

(b) $\displaystyle\sum_{k=1}^{15} (2k-3)$

(d) $\displaystyle\sum_{k=1}^{9} (\frac{1}{k} - \frac{1}{k+1})$.

In Exercises 16 - 19, divide the given interval into n equal subintervals and find the sum of the areas of the n inscribed rectangles for the given function f(x).

16. $[1, 3]$, $n = 4$, $f(x) = x^2 + 1$

(a) Using the command **sum**

(b) Using the command **leftsum** or **rightsum**.

17. $[1, 3]$, $n = 100$, $f(x) = x^2 + 1$

(a) Using the command **sum**

(b) Using the command **leftsum** or **rightsum**.

18. $[0, 5]$, $n = 10$, $f(x) = x^3 + x$

    (a) Using the command **sum**

    (b) Using the command **leftsum** or **rightsum**.

19. $[0, 5]$, $n = 200$, $f(x) = x^3 + x$

    (a) Using the command **sum**

    (b) Using the command **leftsum** or **rightsum**.

20. Use Maple and the First Fundamental Theorem of Calculus to find $\int_1^3 (x^2 + 1)\, dx$. Compare your answer with the results from Exercises 16 and 17.

21. Use Maple and the First Fundamental Theorem of Calculus to find $\int_0^5 (x^3 + x)\, dx$. Compare your answer with the results from Exercises 18 and 19.

22. For $f(x) = x^3 - \frac{9}{2}x^2 + 22$, subdivide $[0, 5]$ into 20 equal subintervals and find the sum of the areas of the inscribed rectangles on the subintervals.

23. For $f(x) = \frac{11}{4}x^2 - \frac{33}{100}x^3 - \frac{9}{2}x + 4$, subdivide $[2, 6]$ into 10 equal subintervals and find the sum of the areas of the inscribed rectangles on the subintervals.

24. For $f(x) = (127\, x^4 - 338\, x^3 - 1268\, x^2 + 3044\, x)/100$, subdivide $[0, R]$ into 100 equal subintervals and find the sum of the areas of the inscribed rectangles on the subintervals. R is the smallest positive root of $f(x) = 0$.

25. Use the First Fundamental Theorem of Calculus to find the area of the finite region between $f(x) = (57\, x^3 - 378\, x^2 + 693\, x - 337)/100$ and the x-axis.

26. Use the First Fundamental Theorem of Calculus to find the area of the region between $f(x) = \sin(2x) + 2\cos(x)$ and the x-axis, from $x = 1$ radian to $x = 5$ radians.

27. Define $f(x) = \dfrac{\ln(x)}{\sqrt{x^3 + 1}}$.

    (a) Is Maple able to find an antiderivative of $f(x)$?

(b) Is Maple able to find the value of the definite integral, $\int_1^3 f(x)\, dx$?

(c) Try evalf(Int(f(x), x = 1..3)); What do you think has taken place to give the value?

(d) Estimate the average value of f(x) on the interval [1, 3].

(e) Estimate the average value of f(x) on the interval [5, 10].

28. Can you find an integral of the following form which Maple cannot compute?

(a) $\int (x - 1)^m (x + 2)^n \, dx$ **(b)** $\int \dfrac{x\, dx}{(x + a)(x^2 + b\,x + c)}$

Projects

1.   For each function f(x) given below, define $F(x) = \int_a^x f(t)\, dt$.

Is F(x) an antiderivative of f(x)? Plot f(x) and F(x) in the same window for $x \in$ [a, b]. How is F(x) related to the area between f(x) and the x-axis on [a, b]?

(a) $f(x) = \dfrac{1}{x^2}$, a = 1, b = 10

(b) $f(x) = \sin x$, a = 0, b = π

(c) $f(x) = \frac{1}{4}x^4 - 2x^3 + 4x^2$, a = 0, b = 5

(d) $f(x) = \frac{1}{4}x^4 - 2x^3 + 4x^2$, a = 0, b = 6.

2.   The integral $\int_0^1 \sqrt{x^3 + 1}\; dx$ represents the area A of a region in the first quadrant.

(a) Try the Maple command for $\int_0^1 \sqrt{x^3 + 1}\; dx$ to find A.

(b) Approximate A by using one of the "sum" commands in the student package (left, middle or right). Which of these three will give the best answer? Use n = 10 and then n = 100. Can you work this problem using the command **sum**?

(c) Can you approximate A to a precision of two decimal places? three decimal places?

(d) To check your answer in part (c), enter the command **evalf(Int(...))**.

3. Estimate the value of the definite integral by using the command **sum** to evaluate the specified Riemann sum:

(a) $\int_{1}^{3} (x^2 + 1)\, dx$, using function values at the right endpoint of each subinterval. For n = 10; for n = 60

(b) $\int_{0}^{2} \sqrt{x^3 + 1}\; dx$, using function values at the midpoint of each subinterval. For n = 10; for n = 80

(c) $\int_{0.5}^{1.75} \sin(x^2)\, dx$, using function values at the point of your choice in each subinterval. For n = 10; for n = 100.

4. Verify your answers in Problem 3 by using the appropriate **leftsum, middlesum,** or **rightsum** command from the student package.

5. Let A represent the area of the region in the first quadrant which is below the graph of $f(x) = x^3 + 2x$, from x = 1 to x = 3. Estimate A by using the command **sum** to evaluate the specified Riemann sum.

(a) Compute the function value at the right endpoint of each subinterval. Take n = 4, 10, and then 50. Finally, take the limit as n approaches infinity.

(b) Compute the function value at the left endpoint of each subinterval. Take n = 4, 10, and then 50. Finally, take the limit as n approaches infinity.

(c) Compute the function value at the midpoint of each subinterval. Take n = 4, 10 and then 50. Finally, take the limit as n approaches infinity.

(d) Note the relationship between these values. How do the values from parts (a), (b), and (c) compare when n = 4? when n = 10? when n = 50? How do the limits compare?

6. Use the commands **leftsum, rightsum,** and **middlesum** to verify your results in Exercise 5, parts (a), (b) and (c).

7.   Define f: $= x^2 + 1$ and let R represent the region under f in the first quadrant from $x = 1$ to $x = 3$.

(a) Use the command **leftbox** to show the four inscribed rectangles which approximate R.

(b) Make a movie showing k inscribed rectangles as k takes on the values 2, 4, 8, 16 and 32.

8.   For each integral below, let $\int_a^b f(x)\, dx = L$. This means that for $\varepsilon > 0$, there exists some positive integer N such that $\left| \sum_{k=1}^{n} f(x_k^*) \Delta x_k - L \right| < \varepsilon$ for $n > N$. Define $\Delta x_k = \dfrac{(b-a)}{n}$ and let $\varepsilon = 0.01$. Taking each x_k^* as the right endpoint of the kth subinterval, find the smallest value of N for which $\left| \sum_{k=1}^{n} f(x_k^*) \Delta x_k - L \right| < \varepsilon$ for $n > N$.

(a) $\displaystyle\int_1^3 (x^2 + 1)\, dx$ (b) $\displaystyle\int_0^{\pi/6} \cos x \, dx$ (c) $\displaystyle\int_{0.5}^{1.75} \sin(x^2)\, dx$.

9.   An experimental method (known as a Monte Carlo process) to estimate the area A of a region R is outlined below.

Enclose R in a rectangular region K. Generate 1000 uniformly distributed random points inside of K and count the number of points S which also fall inside the region R. Then $A \approx \dfrac{S}{1000} * \text{area (K)}$.

Let $K = \{ [x, y],\ 0 \le x \le 2,\ 1 \le y \le 5 \}$. A method to generate the appropriate points is:

```
rn: = rand (0 .. 500)/500.:
s: = 0:
for i from 1 to 1000 do
    xv: = 2 * rn( ):
    yv: = 1 + 4 * rn( )
    if the point [xv, yv] is in R then
            s: = s + 1
    fi:
od:
```

Use the Monte Carlo method to estimate the first quadrant area of the region under f(x), from $x = a$ to $x = b$.

(a) $f(x) = x^2$, $a = 0$, $b = 1$

(b) $f(x) = x^2$, $a = 0$, $b = 2$

(c) $f(x) = \sqrt{x^3 + 1}$, $a = 0$, $b = 1$

(d) $f(x) = \sqrt{x^3 + 1}$, $a = 2$, $b = 4$

(e) $f(x) = \sin(x^2)$, $a = 0.5$, $b = 1.75$.

Verify your answers by finding, or estimating, these areas by a more conventional method.

Chapter 6

Applications of the Definite Integral

New Maple Commands for Chapter 6

exp(x);	exponential function e^x
with(student);	load student package
rightbox(f(x),x=a..b,n);	graphs f(x) between a and b, putting n rectangles under graph of f(x)
rightsum(f(x),x=a..b,n);	exact sum of areas of rectangles from rightbox

---

Introduction

If one views Maple as a "calculator," then the problems in the application section of integration are reduced to button-pushing after formulation. However, we suggest that Maple supports exploratory activities that increase understanding. The examples in this chapter are offered in support of this conviction.

Thus, in Example 3 we want to illustrate computing area between two curves, so we construct the two curves in Examples 1 and 2. Then in Example 4 we obtain the volume of a solid of revolution, and approximate it with a discrete sum in Example 5. We do the same with arc length in Examples 6 and 7.

Examples

Example 1

Find a cubic polynomial f(x) that has a relative maximum at (1, 7) and a relative minimum at (8, -2).

A cubic polynomial is of the form f = a x^3 + b x^2 + c x + d and therefore has four constants (a, b, c, d) to be determined. This requires four conditions leading to four equations. The conditions are that the cubic pass through the points (1, 7) and (8, -2), and that the derivative have zeros at x = 7 and x = 8. Such a "brute-force" technique is implemented in Maple as follows.

- f := a*x^3 + b*x^2 + c*x + d;

$$f := a\,x^3 + b\,x^2 + c\,x + d$$

- e1 := subs(x = 1, f) = 7;
  e2 := subs(x = 8, f) = -2;

$$e1 := a + b + c + d = 7$$

$$e2 := 512\,a + 64\,b + 8\,c + d = \text{-}2$$

- fp := diff(f, x);

$$fp := 3\,a\,x^2 + 2\,b\,x + c$$

- e3 := subs(x = 1, fp) = 0;
  e4 := subs(x = 8, fp) = 0;

$$e3 := 3\,a + 2\,b + c = 0$$

$$e4 := 192\,a + 16\,b + c = 0$$

- q := solve({e1, e2, e3, e4}, {a, b, c, d});

$$q := \left\{ a = \frac{18}{343}, d = \frac{2194}{343}, b = \frac{\text{-}243}{343}, c = \frac{432}{343} \right\}$$

- f1 := subs(q, f);

$$f1 := \frac{18}{343}x^3 - \frac{243}{343}x^2 + \frac{432}{343}x + \frac{2194}{343}$$

Alternatively, we could have started with a representation of the derivative, which perforce would have the factors (x - 1) and (x - 8). An integration, remembering to tack on a constant of integration, then provides f.

- fp := c*(x - 1)*(x - 8);

$$fp := c\,(x-1)\,(x-8)$$

- F := int(fp, x) + d;

$$F := c\left(\frac{1}{3}x^3 - \frac{9}{2}x^2 + 8\,x \right) + d$$

Interpolating the two extreme points with F now leads to just two equations in the two unknowns c and d.

- E1 := subs(x = 1, F) = 7;
  E2 := subs(x = 8, F) = -2;

$$E1 := \frac{23}{6}c + d = 7$$

$$E2 := -\frac{160}{3}c + d = \text{-}2$$

- Q := solve({E1, E2}, {c, d});

$$Q := \left\{ c = \frac{54}{343}, d = \frac{2194}{343} \right\}$$

- subs(Q, F);

$$\frac{18}{343} x^3 - \frac{243}{343} x^2 + \frac{432}{343} x + \frac{2194}{343}$$

The results, not surprisingly, are the same.

Example 2

Find a second cubic g(x) that has a relative minimum at (5, 1) and a relative maximum at (8, 4).

We'll begin with the derivative and integrate to find g(x).

- gp := c*(x - 5)*(x - 8);

$$gp := c (x - 5) (x - 8)$$

- g := int(gp, x) + d;

$$g := c \left(\frac{1}{3} x^3 - \frac{13}{2} x^2 + 40 x \right) + d$$

- e1 := subs(x = 5, g) = 1;
  e2 := subs(x = 8, g) = 4;

$$e1 := \frac{475}{6} c + d = 1$$

$$e2 := \frac{224}{3} c + d = 4$$

- q1 := solve({e1, e2}, {c, d});

$$q1 := \left\{ c = \frac{-2}{3}, d = \frac{484}{9} \right\}$$

- g1 := subs(q1, g);

$$g1 := -\frac{2}{9} x^3 + \frac{13}{3} x^2 - \frac{80}{3} x + \frac{484}{9}$$

Example 3

Find the area of the region bounded by the graphs of f(x) and g(x) found in Examples 1 and 2.

Begin with a plot showing both f1 and g1, the names assigned to f(x) and g(x), respectively.

• plot({f1, g1}, x = 3..10);

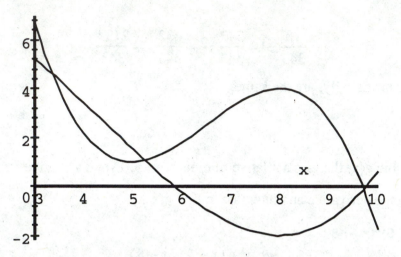

Next, compute the points of intersection between f1 and g1. The equation f1 = g1 is a cubic equation and can be solved exactly with Maple's **solve** command. However, we opt for the simplicity of the results of the numeric solution provided by **fsolve**.

• r := fsolve(f1 = g1, x);

$$r := 3.374774810, 5.257242355, 9.721756420$$

The algorithm for finding the area between two curves requires that we distinguish between the "upper" curve and the "lower" curve.

• q2 := Int(f1 - g1, x = r[1]..r[2]) + Int(g1 - f1, x = r[2]..r[3]);

$$q2 := \int_{3.374774810}^{5.257242355} \frac{848}{3087}x^3 - \frac{5188}{1029}x^2 + \frac{28736}{1029}x - \frac{146266}{3087}\,dx$$

$$+ \int_{5.257242355}^{9.721756420} -\frac{848}{3087}x^3 + \frac{5188}{1029}x^2 - \frac{28736}{1029}x + \frac{146266}{3087}\,dx$$

• value(q2);

$$18.4147532$$

Example 4

A solid is formed by revolving the curve $y = \sqrt{(x^3 + 1)}$ about the x-axis (for $0 \le x \le 2$). Find the volume of the solid of revolution so formed.

The volume v is given by the integral $\pi \int r^2\, dx$, provided the radius r is given by

• f := sqrt(x^3 + 1);

$$f := \sqrt{x^3 + 1}$$

• v := Pi * Int(f^2, x = 0..2);

$$v := \pi \int_0^2 x^3 + 1\, dx$$

• value(v);

$$6\,\pi$$

Example 5

Approximate the volume in Example 4 by summing the volumes of 100 approximating disks whose radii are the values of f at the right-hand endpoints of subintervals formed on the x-axis.

A sketch generated by **rightbox** from the student package will help visualize this computation.

• with(student):
• rightbox(f, x = 0..2, 10);

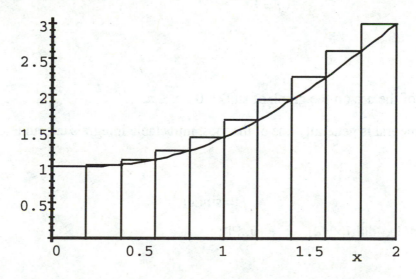

If this cross section were rotated about the x-axis, the rectangles shown would generate the approximating disks whose volumes we need to sum. For 100 disks, the thickness of each disk would be 2/100 and the x-coordinate at the right-hand edge of the kth subinterval would be $x_k = k/50$. The approximating sum would be

• q3 := (Pi/50)*Sum(subs(x = k/50, f)^2, k = 1..100);

$$q3 := \frac{1}{50}\pi\left(\sum_{k=1}^{100}\left(\frac{1}{125000}k^3 + 1\right)\right)$$

• value(q3);
  evalf(q3);

$$\frac{15201}{2500}\pi$$

$$19.10213997$$

Incidentally, the sum in q3 is nothing more than what the **rightsum** command would have created.

• rightsum(Pi*f^2, x = 0..2, 100);

$$\frac{1}{50}\left(\sum_{i=1}^{100}\pi\left(\frac{1}{125000}i^3 + 1\right)\right)$$

Example 6

Find the length of the arc on the graph of sin(x), $0 \le x \le \pi$.

The arc length integral is generally one of the most intractable integrals encountered in the calculus.

• f := sin(x);

$$f := \sin(x)$$

• q4 := int(sqrt(1+ diff(f,x)^2), x = 0..Pi);

$$q4 := \int_0^\pi \sqrt{1 + \cos(x)^2}\, dx$$

By returning the integral unevaluated, even though we used the **int** operator, Maple indicates that it could not find an appropriate antiderivative. We then evaluate this integral numerically via

• evalf(q4);

3.820197789

Example 7

Approximate the arc length of Example 6 as a discrete sum of the lengths of edges in an approximating polygon. Take a polygon with 100 edges.

Perhaps a sketch will help.

• plot({f, [seq([k*Pi/3, subs(x = k*Pi/3, f)], k = 0..3)]}, x = 0..Pi);

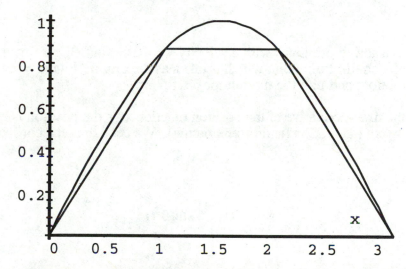

The lengths of line segments must be computed and summed. First, unassign the counter k, and then create a function p(k) that returns the partition points $k\pi/100$. The lengths of the polygonal line segments are nothing more than $\sqrt{[\Delta x]^2 + [\Delta y]^2}$.

• k := 'k';
  p := k -> k*Pi/100;
  dk := sqrt((p(k+1) - p(k))^2 + (sin(p(k+1)) - sin(p(k)))^2);

q5 := Sum(dk, k = 0..99);

$$k := k$$

$$p := k \rightarrow \frac{1}{100} k \pi$$

$$dk := \sqrt{\left(\frac{1}{100}(k+1)\pi - \frac{1}{100}k\pi\right)^2 + \left(\sin\left(\frac{1}{100}(k+1)\pi\right) - \sin\left(\frac{1}{100}k\pi\right)\right)^2}$$

$$q5 := \sum_{k=0}^{99} \sqrt{\left(\frac{1}{100}(k+1)\pi - \frac{1}{100}k\pi\right)^2 + \left(\sin\left(\frac{1}{100}(k+1)\pi\right) - \sin\left(\frac{1}{100}k\pi\right)\right)^2}$$

- evalf(q5);

$$3.820148519$$

The figure suggested that a polygonal line would approximate sin(x) very closely.

Example 8

A mass attached to a spring oscillates with a velocity $v = 10\ e^{-t} \sin(3t)$. At time $t = 0$, the mass was 2 feet below the equilibrium point which is taken as the origin. Find $x = x(t)$, the position function for this motion, and plot the motion for $0 \le t \le 3\pi$.

Since velocity is the time-derivative of the position function x(t), the position is recoverable from the velocity by an integration (antidifferentiation). We need to add a constant of integration!

- v   :=   10*exp(-t)*sin(3*t);

$$v := 10\ e^{(-t)} \sin(3\ t)$$

- q6 := int(v, t) + c;

$$q6 := -3\ e^{(-t)} \cos(3\ t) - e^{(-t)} \sin(3\ t) + c$$

The constant of integration is evaluated by imposing the condition that the initial position was -2 (as in "2 feet below the origin").

- q7 := subs(t = 0, X) = -2;

$$q7 := -3\,e^0 \cos(0) - e^0 \sin(0) + c = -2$$

- solve(q7, c);

$$1$$

- xt := subs(c = 1, q6);

$$xt := -3\,e^{(-t)} \cos(3\,t) - e^{(-t)} \sin(3\,t) + 1$$

- plot(xt, t = 0..3*Pi);

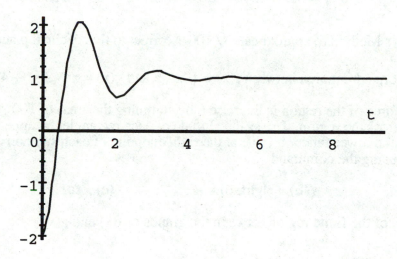

Exercises

1. Find the coefficient of x^2 in a cubic polynomial $f(x)$ having roots $x = -2$, $x = 3$ and $x = 5$.

2. Find the cubic polynomial $g(x)$ having critical points at $(-2, 2)$ and $(1.5, -1)$.

3. Plot the graphs of $f(x)$ and $g(x)$ from Exercises 1 and 2 on the same coordinate axes.

4. Where is the graph of $y = 33x^4 + 80x^3 - 649x^2 + 23.96x - 3594$ concave down?

5. Use Newton's Method to approximate $\sqrt[3]{1001}$ correct to five decimal places.

6. Find the first quadrant area between $f(x) = 10x - 5$ and $g(x) = x^3 - 2x^2 + 4x - 1$.

7. Estimate the area of the region in Exercise 6 by summing the areas of 100 "thin" rectangles across the region. Use as the altitude of each rectangle the upper function value minus the lower function value at the right endpoint of each subinterval. Carry out this process using the command

    (a) sum   **(b) rightsum**   **(c) for**.

8. Find the area of the finite region between the graphs of $f(x)$ and $g(x)$.

    (a) $f(x) = x^6 - 6x + 11$, $g(x) = 10x - x^2 - 15$

    (b) $f(x) = -\frac{1}{3}x^3 + \frac{7}{2}x^2 - 10x + 13$, $g(x) = \frac{1}{3}x^3 - 3x^2 + 8x$

    (c) $f(x) = 2x^3 - 21x^2 + 36x + 108$, $g(x) = -8x^3 + 108x^2 - 336x + 179$.

9. Find the volume when the region between $y = 0$, $y = \dfrac{3}{\sqrt{3x - 0.5}}$, $x = 1.2$ and $x = 1.9$ is revolved about the x-axis.

10. A solid is formed by revolving $y = x^3$ about the x-axis (for $0 \le x \le 4$). Use the Maple command **sum** to approximate the volume of the solid by summing the volumes of 100 circumscribed right-circular cylinders with equal heights.

11. By integration, find the volume of the solid in Exercise 10. Compare the results from Exercises 10 and 11.

12. Repeat Exercise 10 with 100 circumscribed right-circular shells having equal base thicknesses. Use the **rightsum** command.

13. Use shells and integration to find the volume of the solid in Exercise 10. Compare the results from Exercises 12 and 13.

14. Approximate the arc length of the curve $y = \sqrt{x}$ from $x = 1$ to $x = 4$ by summing the lengths of 100 appropriate hypotenuses. (Work on equal subintervals of [1, 4].)

15. An object moves along a straight line with acceleration $a = 40\,t^3 - 243\,t^2 + 154\,t + 355$ ft/sec^2. At time $t = 0$, it is moving (to the left) with velocity $v = -35.7$ ft/sec^2.

    (a) At what times is the object stopped?

    (b) What is its velocity when it is first 16 feet from the origin? (Assume that it is 10 feet from the origin when $t = 0$.)

16. A mass is attached to a spring which in turn is attached to a support which oscillates in a way that imparts the following velocity to the mass:

    $$v = 43 \sin t + 25 \cos t + 25 \exp(-2t)\,(\sin(2t) - \cos(2t)).$$

    Given that the mass is 9 feet above the origin at time $t = 0$, find the position of the mass at time t, and plot the motion of the mass for $0 \leq t \leq 4\pi$.

17. An object moves along a straight line with velocity $v = 324 - 391t + 100\,t^2$ ft/sec^2. If its initial position is 10 feet to the left of the origin, what is the total distance traveled by the object during the first 4 seconds?

18. A tank is formed by revolving the curve $y = \sqrt{x^3 - 8}$ about the y-axis ($2 \leq x \leq 5$). If the tank is filled with water to a depth of 5 feet, find (approximately) the amount of work required to pump the water out of the top of the tank.

Projects

1. A non-negative function f(x) is to pass through the points (0, 4) and (4, 2). The area A between f(x) and the x-axis from $x = 0$ to $x = 4$ is to have the value specified. Find a function that meets these requirements. Is your answer unique?

    (a) $A = 15$ **(b)** $A = 10$.

2. Estimate the area of the finite region between $f(x) = x^2/8 - x$ and $g(x) = x/2$ by summing the areas of n inscribed rectangles. Use equal bases, being careful of the rectangle with the lowest base.

    (a) $n = 100$ **(b)** $n = 500$.

3. Estimate the area of the finite region between the curves defined implicitly by $8x = y^2 - 16y + 48$ and $8x = 8y - y^2 + 16$, by summing the areas of n circumscribed rectangles. Use equal bases. See the command **implicit** in the plots package.

   (a) n = 100   (b) n = 500.

4. (a) A solid is formed by revolving about the x-axis, the function defined by $f(x) = \sqrt{x^3 + 1}$, $0 \le x \le 2$. Estimate the volume of this solid by summing the volumes of 100 right-circular cylinders of equal height. Use the function value at the right endpoint of each subinterval as the radius of each cylinder.

   (b) Verify your answer to part (a) by applying the command **evalf** to the appropriate definite integral.

5. In this problem you can show that a certain bucket full of paint does not hold enough paint to cover a cross-sectional portion of the inside of the bucket.

   (a) Find the area A under the curve $y = \dfrac{1}{x}$ from x = 1 to x = a (a > 1).

   (b) Find the volume V of the solid formed by revolving the curve $y = \dfrac{1}{x}$ $(1 \le x \le a)$ about the x=axis.

   (c) Show that V < A when a is large enough.

6. Compelled to fence my dog, I have scrounged some lengths of picket fence from the dump. The four segments I found have lengths of 3, 4, 5, and 6 meters. If I form a quadrilateral using these segments, what is the maximum area that I can fence?
   -Technology Review (The Bent)

7. From a circular disk, a sector is to be removed and the remaining material joined along the cut radii to form a cone. What central angle sector should be removed to have maximum cone volume?
   - E.K. Key (The Bent)

8. In this project you will approximate the arc length of a curve by using straight-line segments. These segments are actually secant lines, and the x-coordinates of the end points of the secant lines are the uniformly spaced values $0, \dfrac{\pi}{5}, \dfrac{2\pi}{5}$, etc.

   (a) Obtain a graph of sin (x) and its approximation via five secant lines.

   (b) Find the sum of the lengths of the five secant lines in part (a). Compare this sum with the arclength of sin (x) over [0, π].

   (c) Repeat part (b) using 100 secant lines.

Chapter 7

Logarithmic and Exponential Functions

New Maple Commands for Chapter 7

arcsinh(x)	function inverse to sinh(x)
assume(x>0);	have Maple impose restriction $x \geq 0$
convert(sinh(x),exp);	change sinh(x) to $[e^x - e^{-x}]/2$
convert(exp(x),trig);	change exp(x) to $\cosh(x) + \sinh(x)$
dsolve({diff(y(t),t)=1,y(0)=2},y(t));	solve differential equation $y'(t) = 1$, with initial condition $y(0) = 2$, for y(t)
exp(x);	exponential function e^x
ln(x);	natural logarithm, $\log_e x$
log[2](x);	base-2 logarithm of x, $\log_2 x$

Introduction

The two functions e^x and ln x that appear regularly in the applications of calculus are related to each other as inverse functions. Hence, we will spend a moment recalling the notion of an inverse function, and the algorithm by which inverse functions are computed, before we examine the exponential and log functions.

If the graph of a function f is defined as a collection of ordered pairs (x, f(x)), then the graph of g, the inverse of f, is defined as the collection of ordered pairs (f(x), x). Thus, reversing the position of the ordinate and abscissa for the function generates the graph of the inverse. Hence, the graphs of f and its inverse are reflections of each other across the line y = x.

Operationally, two functions f and g are declared to be related as inverses if f(g(x)) = g(f(x)) = x. Algorithmically, the rule (formula) for g(x), the inverse of f, is computed by solving the equation f(x) = y for x, and then switching the letters x and y. We illustrate by examples.

Examples

Example 1

On one set of axes, plot $f(x) = 1 + x^2$, $x \geq 0$, the line y = x, and the graph of the function whose ordered pairs are the reverse of those for f(x).

To plot the graph of these reversed ordered pairs, use a parametric representation.

• f := 1 + x^2;

$$f := 1 + x^2$$

- plot({[x, f, x = 0..3], [x, x, x = 0..8], [f, x, x = 0..3]}, scaling = constrained);

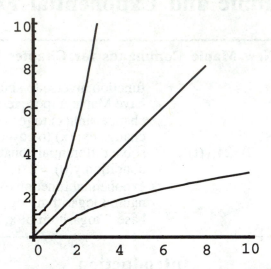

The graphs seem to be mirror images across the line y = x. To verify that we have graphed the inverse of f, we need to apply the algorithm for computing an inverse.

Example 2

Compute the formula for the function inverse to f(x) given in Example 1.

- q := solve(f = y, x);

$$q := \sqrt{-1 + y}, -\sqrt{-1 + y}$$

The step "solve the equation f(x) = y for x" resulted in two possible expressions for x. Since the domain of f(x) is the set {x | x ≥ 0}, we need to pick the first branch in q, that is, the one that is always positive. Switch the letters in this expression to form the rule for g(x), the inverse of f(x).

- g := subs(y = x, q[1]);

$$g := \sqrt{-1 + x}$$

Example 3

Verify that f and g from Examples 1 and 2 are inverses by showing that f(g(x)) = g(f(x)) = x.

- q1 := subs(x = g, f);

$$q1 := x$$

- q2 := subs(x = f, g);

$$q2 := \sqrt{x^2}$$

Maple will not simplify q2 to x because such a transformation is only correct if $x \geq 0$. While we know that this is so, Maple does not, and hence will not carry out the simplification.

- assume(x>0);
- simplify(subs(x = f, g));

$$x\sim$$

The tilde (~) attached to the x is a reminder that the result just produced is correct only in light of the assumption made about x. To remove the assumption on x, use the same technique as for removing values assigned to a variable.

- x := 'x';

$$x := x$$

Example 4

Find an inverse for $f = x^3 + 4\,x^2 + 12$, if one exists. If not, determine the largest interval containing $x = 1$ over which f will have an inverse, and then determine that inverse.

A function is invertible only where it is one-to-one, so invertible functions are strictly monotone (either increasing or decreasing). One can determine monotonicity either from a graph or from the behavior of the first derivative, which is positive for a monotone increasing function and negative for a monotone decreasing one.

- f := x^3 + 4*x^2 + 12;

$$f := x^3 + 4\,x^2 + 12$$

- plot(f, x = -5..2);

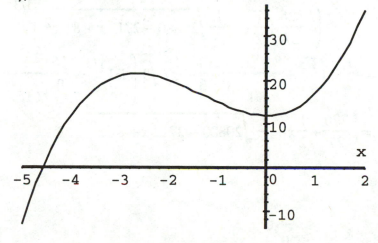

This function is not monotone and hence does not have an inverse. However, $x = 1$ is in an interval over which f is increasing. This interval can be found via

- solve(diff(f, x) > 0, x);

$$\left\{x < \frac{-8}{3}\right\}, \{0 < x\}$$

We therefore invert f on the interval $x \geq 0$. The algorithm for computing the inverse function is "solve the equation $y = f(x)$ for x, then switch the letters."

- q3 := solve(f = y, x);

$$q3 := \%1^{1/3} + \frac{16}{9}\frac{1}{\%1^{1/3}} - \frac{4}{3}, -\frac{1}{2}\%1^{1/3} - \frac{8}{9}\frac{1}{\%1^{1/3}} - \frac{4}{3} + \frac{1}{2}I\sqrt{3}\left(\%1^{1/3} - \frac{16}{9}\frac{1}{\%1^{1/3}}\right),$$

$$-\frac{1}{2}\%1^{1/3} - \frac{8}{9}\frac{1}{\%1^{1/3}} - \frac{4}{3} - \frac{1}{2}I\sqrt{3}\left(\%1^{1/3} - \frac{16}{9}\frac{1}{\%1^{1/3}}\right)$$

$$\%1 := -\frac{226}{27} + \frac{1}{2}y + \frac{1}{18}\sqrt{20880 - 2712\,y + 81\,y^2}$$

There are three branches to examine, but first, let's finish the inversion algorithm by switching letters and then naming each branch separately.

- for k from 1 to 3 do g.k := subs(y = x, q3[k]); od;

$$g1 := \left(-\frac{226}{27} + \frac{1}{2}x + \frac{1}{18}\sqrt{20880 - 2712\,x + 81\,x^2}\right)^{1/3}$$

$$+ \frac{16}{9}\frac{1}{\left(-\frac{226}{27} + \frac{1}{2}x + \frac{1}{18}\sqrt{20880 - 2712\,x + 81\,x^2}\right)^{1/3}} - \frac{4}{3}$$

$$g2 := -\frac{1}{2}\%1^{1/3} - \frac{8}{9}\frac{1}{\%1^{1/3}} - \frac{4}{3} + \frac{1}{2}I\sqrt{3}\left(\%1^{1/3} - \frac{16}{9}\frac{1}{\%1^{1/3}}\right)$$

$$\%1 := -\frac{226}{27} + \frac{1}{2}x + \frac{1}{18}\sqrt{20880 - 2712\,x + 81\,x^2}$$

$$g3 := -\frac{1}{2}\%1^{1/3} - \frac{8}{9}\frac{1}{\%1^{1/3}} - \frac{4}{3} - \frac{1}{2}I\sqrt{3}\left(\%1^{1/3} - \frac{16}{9}\frac{1}{\%1^{1/3}}\right)$$

$$\%1 := -\frac{226}{27} + \frac{1}{2}x + \frac{1}{18}\sqrt{20880 - 2712\,x + 81\,x^2}$$

At this point, a graph of f and its inverse on $x \geq 0$ is very helpful.

- plot({[x, f, x = 0..2], [f, x, x = 0..2]}, scaling = constrained);

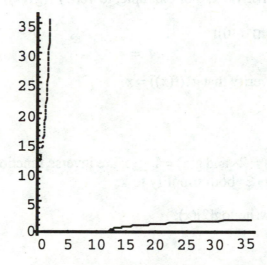

The graph tells us that the inverse function has domain $x \geq 12$, and we seek the branch that is real and positive for $x \geq 12$. A table of values for each of g1, g2, and g3 is now revealing.

- for k from 12 to 20 do
     k, evalf(subs(x = k, [g1, g2, g3]), 5);
  od;

$$12, [0, -4.0001 - .00003\,I, .0001 - .00003\,I]$$

$$13, [.4729 - .00001\,I, -3.9355 - .00001\,I, -.5373 - .00001\,I]$$

$$14, [.6555 - .00001\,I, -3.8663 - .00001\,I, -.7891 - .00001\,I]$$

$$15, [.7913 - .00001\,I, -3.7914, -.9998]$$

$$16, [.9033, -3.7093, -1.1939]$$

$$17, [1.0001, -3.6181, -1.3819]$$

$$18, [1.0862 - .00001\ I, -3.5141 - .00009\ I, -1.5719 + .00009\ I]$$

$$19, [1.1643 - .00001\ I, -3.3914, -1.7728]$$

$$20, [1.2361, -3.2361, -2.0000]$$

Branch g1 satisfies g1(12) = 0, but for some values of x (e.g., x = 14) g1 evaluates to a complex number with a small imaginary part. This is spurious, a function of round-off error. The branch we want is g1, but a plot of g1 would fail because of the way Maple treats floating-point calculations. To verify that g1 is the appropriate inverse we must show it satisfies the relationships f(g1(x)) = g1(f(x)) = x. For example, to verify f(g1(x)) = x,

- q4 := simplify(subs(x = g1, f));

$$q4 := x$$

We leave it to the reader to verify that g1(f(x)) = x.

Example 5

Show that the functions $f(x) = 2^x$ and $g(x) = \log_2 x$ are inverse functions by showing that the compositions $2^{\log_2 x}$ and $\log_2 2^x$ both simplify to x.

To enter $\log_2 x$ into Maple write log[2](x).

- simplify(2^log[2](x));

$$x$$

- simplify(log[2](2^x));

$$x$$

Example 6

Show that the functions $f(x) = e^x$ and $g(x) = \ln x$ are inverse functions by showing that the compositions $e^{\ln x}$ and $\ln e^x$ both simplify to x.

To enter ln x into Maple, write ln(x) and to enter e^x, write exp(x).

- simplify(exp(ln(x)));

$$x$$

- simplify(ln(exp(x)));

$$x$$

Example 7

Show that e is the limit, as r approaches infinity, of $(1 + 1/r)^r$. Then evaluate e as a floating-point number. Show that while Maple uses e as output, exp(1) is the appropriate input for the constant e.

- limit((1 + 1/r)^r, r = infinity);

$$e$$

- evalf(e, 15);

$$e$$

- evalf(exp(1), 15);

$$2.71828182845905$$

Example 8

Find, on the interval $0 \le x \le 2\pi$, the absolute maximum and the absolute minimum of the function $f(x) = e^{\left(\frac{1}{\sin(x) + \sec(x)}\right)}$.

- f := exp(1/(sin(x) + sec(x)));

$$f := e^{\left(\dfrac{1}{\sin(x) + \sec(x)}\right)}$$

- plot(f, x = 0..2*Pi);

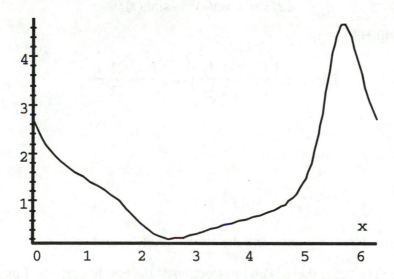

- fp := diff(f, x);

$$fp := -\frac{(\cos(x) + \sec(x)\tan(x))\, e^{\left(\frac{1}{\sin(x) + \sec(x)}\right)}}{(\sin(x) + \sec(x))^2}$$

- x1 := fsolve(fp = 0, x, 2..3);
  x2 := fsolve(fp = 0, x, 5..6);

$$x1 := 2.542825948$$

$$x2 := 5.684418602$$

- y1 := evalf(subs(x = x1, f));
  y2 := evalf(subs(x = x2, f));

$$y1 := .2131766413$$

$$y2 := 4.690945475$$

From these computations and the graph of f(x), we conclude that the absolute minimum is at (x1, y1) while the absolute maximum is at (x2, y2).

Example 9

Verify the fundamental identity for hyperbolic functions, $\cosh^2(t) - \sinh^2(t) = 1$. (<u>Hint</u>: Do it the *eeezy* way by converting each hyperbolic function to its exponential equivalent.)

- LHS := cosh(t)^2 - sinh(t)^2;

$$LHS := \cosh(t)^2 - \sinh(t)^2$$

- LHS1 := convert(LHS, exp);

$$LHS1 := \left(\frac{1}{2}e^t + \frac{1}{2}\frac{1}{e^t}\right)^2 - \left(\frac{1}{2}e^t - \frac{1}{2}\frac{1}{e^t}\right)^2$$

- simplify(LHS1);

$$1$$

Example 10

Find the area enclosed by the y-axis, and the curves sinh(2x) and cosh(x).

First, look at a graph to see if there is an intersection of the two hyperbolic functions.

- plot({sinh(2*x), cosh(x)}, x = 0..1);

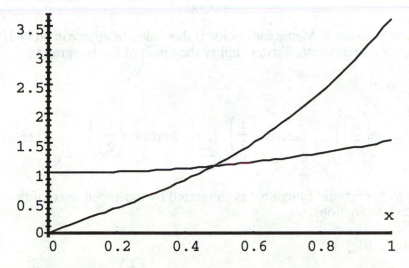

Next, compute the point of intersection; this will give the rightmost limit of integration.

- solve(sinh(2*x) = cosh(x), x);

$$\frac{1}{2} I \pi, \operatorname{arcsinh}\left(\frac{1}{2}\right)$$

- evalf(arcsinh(1/2));

$$.4812118251$$

- q5 := Int(cosh(x) - sinh(2*x), x = 0..arcsinh(1/2));

$$q5 := \int_{0}^{\operatorname{arcsinh}\left(\frac{1}{2}\right)} \cosh(x) - \sinh(2\,x)\,dx$$

- q6 := value(q5);

$$q6 := -\frac{1}{4}\frac{2\,\%2^2\,\%1 - 2\,\%1 + \%2\,\%1^2 + \%2}{\%2\,\%1} + \frac{1}{2}$$

$$\%1 := e^{\left(-2\,\operatorname{arcsinh}\left(\frac{1}{2}\right)\right)}$$

$$\%2 := e^{\left(-\operatorname{arcsinh}\left(\frac{1}{2}\right)\right)}$$

- evalf(q6);

$$.2500000001$$

It might be instructive to see if Maple can decide if this value of approximately 1/4 is symbolically exactly that amount. First, simplify the result of the integration.

- q7 := simplify(q6);

$$q7 := -\frac{1}{2}e^{\left(-\text{arcsinh}\left(\frac{1}{2}\right)\right)} + \frac{1}{2}e^{\left(\text{arcsinh}\left(\frac{1}{2}\right)\right)} - \frac{1}{4}e^{\left(-2\,\text{arcsinh}\left(\frac{1}{2}\right)\right)} - \frac{1}{4}e^{\left(2\,\text{arcsinh}\left(\frac{1}{2}\right)\right)} + \frac{1}{2}$$

Then, convert q7 to hyperbolic functions, as suggested by the appearance of the exponentials in the appropriate combinations.

- q8 := convert(q7, trig);

$$q8 := 1 - \frac{1}{2}\cosh\left(2\,\text{arcsinh}\left(\frac{1}{2}\right)\right)$$

Finally, experience teaches that an **expand** will clear the parentheses.

- expand(q8);

$$\frac{1}{4}$$

Example 11

The relative rate of decay of radon gas is a constant, and 10% of an initial volume of the gas will decay in 14.5 hours. How long will it take for this gas to decay to 10% of the initial volume?

First, we formulate the model of this decay process as a differential equation.

- q9 := diff(x(t),t)/x(t) = k;

$$q9 := \frac{\dfrac{\partial}{\partial t}x(t)}{x(t)} = k$$

Although equation q9 is algebraically equivalent to the equation x'(t) = k x(t), the verbal description "the rate of decay is proportional to the amount of gas present" that this formulation inspires does not make sense physically. The implied question would be, "How do the molecules of the gas know that they should adjust their decay rate to accommodate a differing sample size?" Viewing the process as one in which the rate of decay per unit volume is constant avoids anthropomorphizing radon gas.

Solving the differential equation with the initial condition that $x(0) = x_0$ leads to

- q10 := dsolve({q9, x(0) = x0}, x(t));

$$q10 := x(t) = e^{(k\,t)}\,x0$$

Applying the condition that at time $t = 14.5$ the volume $x(14.5)$ is $.9\,x_0$, we get

- q11 := subs({x(t) = 9*x0/10, t = 29/2}, q10);

$$q11 := \frac{9}{10}x0 = e^{\left(\frac{29}{2}k\right)}x0$$

- q12 := solve(q11, k);

$$q12 := \frac{2}{29}\ln\left(\frac{9}{10}\right)$$

- q13 := subs(k = q12, q10);

$$q13 := x(t) = e^{\left(\frac{2}{29}\ln\left(\frac{9}{10}\right)t\right)}x0$$

Equation q13 is a model for the decay of this sample of gas. We are ready to ask of this model the question of interest, namely, "How long will it take for the amount of radon gas to be reduced to $x_0/10$?"

- q14 := subs(x(t) = x0/10, q13);

$$q14 := \frac{1}{10}x0 = e^{\left(\frac{2}{29}\ln\left(\frac{9}{10}\right)t\right)}x0$$

- q15 := solve(q14, t);

$$q15 := -\frac{29}{2}\frac{\ln(10)}{\ln\left(\frac{9}{10}\right)}$$

• evalf(q15);
  evalf(q15/24);

$$316.8880071$$

$$13.20366696$$

The first answer is in hours; the second, in days.

Exercises

1. Show that g(x), the function inverse to f(x), exists. Find g(x), evaluate f(g(2)) and g(f(2)), and verify that f(g(x)) = x = g(f(x)).

    (a) $f(x) = x^3 + 3 x^2 + 3 x + 3$ (b) $f(x) = x^3 + 5 x$.

2. Determine the largest interval containing x = a on which f(x) has an inverse.

    (a) $f(x) = 3 x^4 - 4 x^3 - 12 x^2 + 13$, a = 1

    (b) $f(x) = \sin(x)$, a = 1.57

    (c) $f(x) = \sin(x^2 + 0.1)$, a = 1.2

    (d) $f(x) = \sin(x^2 + 0.1)$, a = 1.3.

3. Plot the graphs of $f(x) = \log_{10} x$ and g(x) = ln x in the same window for $0.1 \le x \le 12$. (See ? log.)

4. Plot the graphs of $f(x) = 3^x$ and $g(x) = 2^x$ in the same window for $-1 \le x \le 2$.

5. Use Maple to evaluate

    (a) $\int_0^4 \frac{dx}{x + 4}$ (b) $\int_1^2 \frac{dx}{5x - 3}$.

6. Approximate ln 3 by summing the areas of 50 rectangles having bases on [1, 3], using $f(t) = \frac{1}{t}$ and

    (a) inscribed rectangles   (b) circumscribed rectangles.

7. Find the minimum value of f(x). Be sure to create a plot.

    (a) $f(x) = x^3 + x - 2 \ln x$ (c) $f(x) = (x^3 - 2 x) \ln(2 x) + 4$.

    (b) $f(x) = (x^3 + x) \ln(2x)$

8. Find an equation of the line tangent to the graph of $y = \ln((x^2 - 4)^3)$ at the point where $x = \sqrt{5}$.

9. Find the value of

    (a) exp (0)   (b) exp (1).

10. Find the value of

    (a) exp (2 ln 3)   (c) exp (4 ln 2)

    (b) $\exp(\frac{3}{8})$ (d) $4^{1.5}$.

11. Find the value of

    (a) $\log_2 10$ (b) $\log_3 2$.

12. At time $t = 0$ a particle is 7.5 units to the right of the origin, and moving to the right with a velocity $v = 20 - e^{t/2}$. Plot the velocity function and the position function in the same window for $0 \leq t \leq 5$. At what time is the particle 50 units to the right of the origin?

13. Plot the graph of $f(x) = e^{x/10}(\sin x)$ for $-15 \leq x \leq 15$. From the graph, can you predict the value of each limit?

    (a) $\lim_{x \to -\infty} f(x)$ (c) $\lim_{x \to -\infty} e^{x/m}(\sin x)$

    (b) $\lim_{x \to +\infty} f(x)$ (d) $\lim_{x \to +\infty} e^{x/m}(\sin x)$ for large positive m.

14. Define $f(x) = \dfrac{\ln (x^2 + 1)}{x}$. What is the largest possible domain of $f(x)$? Will the graph of $f(x)$ have any symmetries? Plot the graph of $f(x)$ and then find the maximum and minimum values of $f(x)$.

15. Define $f(x) = x^3 e^{-x}$. Predict the behavior of $f(x)$ as x becomes large. Find $\lim_{x \to +\infty} f(x)$ and plot $f(x)$.

16. Plot $f(x) = x^3$ and $g(x) = e^x$ in the same window. Find <u>all</u> intersection points of $f(x)$ and $g(x)$.

17. Find the maximum value of $f(x) = (e^{x/2} - 1) \sin 3x$ on $[0, 10]$.

18. Find an equation of the line tangent to the graph of $f(x) = \cosh x$ at the point where $x = 3$.

19. Find the inverse of the function $f(x) = \ln x$.

20. Solve the intial-value problem $\frac{dy}{dx} = -x\,y;\ y(0) = 4$.

21. **(a)** If the purchasing power of the dollar declines at the rate of 4% annually, how long does it take for the purchasing power to decline to half its present value? To what value will the dollar drop after 20 years?

(b) Answer these questions if the rate of decline is 10%.

22. Newton's law of cooling states that the rate at which the temperature T(t) changes in a cooling body is proportional to the difference between the temperature in the body and T_0, the constant temperature of the surrounding medium. Thus, for some constant k,

$$\frac{dT}{dt} = k(T - T_0).$$

A large loaf of banana bread is removed from the oven, and its temperature is recorded as 370°; after 5 minutes, its temperature is 310°. If the room temperature is kept at 68°, how long does it take for the bread temperature to reach 100°? What is the temperature of the bread at the end of 2 hours?

Projects

1. This project examines the exponential function $f(x) = x^x$.

(a) Execute the command "plot (x^x, x = -1 .. 1);". Why does Maple give a partial plot? (It is reasonable to hope that if $x_1 \approx x_2$, then $f(x_1) \approx f(x_2)$. Find, if possible, $f(-\frac{1}{10})$ and $f(-\frac{1}{11})$.)

(b) Is a point at $x = 0$ plotted? Evaluate $f(0)$.

(c) Can you find (or estimate) $\lim\limits_{x \to 0^+} f(x)$?

(d) Estimate the value of each expression and then evaluate using Maple:

$$f(0.1),\ f(0.01),\ f(-1),\ f(-0.5),\ f(-0.4),\ f(-\tfrac{1}{3}),\ f(-\tfrac{1}{4}).$$

Explain the appearance of complex numbers (containing I) in some of these results.

2. Four points on the graph of $f(x) = \ln(x)$ are given: (1, 0), (3, 1.099), (6, 1.792), and (10, 2.303).

(a) Find a cubic polynomial pinv(x) which approximates the inverse function of f(x) by fitting the cubic pinv(x) to the points (0, 1), (1.099, 3), (1.792, 6), (2.303, 10).

(b) As you may know, the inverse function of ln(x) is exp(x). Compose the values of pinv(x) and ln(x) at x = 1, 1.1, 2, 2.3, 3 and 4.

(c) Plot the graphs of pinv(x) and exp(x) in the same window for -1 ≤ x ≤ 3. Why are the graphs close to each other for some values of x but not for others?

3.   An effective (but somewhat abstract) means to define the function ln x is

$$\ln x = \int_1^x \frac{1}{t}\, dt \text{ for } x > 0.$$

(a) Generate a plot of $\frac{1}{t}$ for $1 \le t \le 10$. Use the command **rightsum** to estimate ln 2. Use n = 4, then n = 10. Use the command **rightbox** to look at your approximations. Are your approximations too large or too small?

(b) Repeat part (a) for ln 9.

(c) Can you find a more accurate approximation to ln 2 using the command **middlesum**, with n = 4 and then n = 10?

4.   Define $g(x) = \int_{t=1}^{t=x} \frac{1}{t}\, dt$.

(a) Evaluate g(x) for x = 2, 4, 6, 8, 10. Plot the five points on g(x) which are determined by these five values of x.

(b) Display plots of g(x) and ln x in the same window for 1 ≤ x ≤ 11.

(c) Repeat part (b) for g(x) and h(x) = 0.1 + ln x.

(d) You should notice that g(x) behaves like a logarithm function, although it has been defined as an integral.

5.   The half-life of a certain anti-inflammatory drug is approximately two days. Assuming that the drug's plasma concentrations peak in six hours, plot the amount of drug in the system for 0 ≤ t ≤ 28 days if

(a) one capsule is taken every day.

(b) one capsule is taken every third day.

6.   My neighbor won't tell me when his house mortgage will be paid, but he did reveal the following. He just made his nineteenth monthly payment, consisting of $238.58

principal and $461.44 interest. The previous payment was $237.00 principal and $463.02 interest. What balance does he now owe, and how many more months of payments will he have? Assume a fixed interest rate.

- Byron R. Adams (<u>The Bent</u>)

7.   I put up a clothesline between the tops of two poles that are 2 meters high and 6 meters apart. If the length of the line is 6.06 meters, by how much does it clear the floor?

- Oscar Lanzi III (<u>The Bent</u>)

8.   Define $f(x) = 2 \cosh x$.

(a) Plot $f(x)$ and exp x in the same window for $-1 \leq x \leq 3$.

(b) Find an equation of a parabola $g(x) = ax^2 + bx + c$ which passes through the points$(0, f(0))$, $(0.8, f(0.8))$ and $(1.5, f(1.5))$. Plot $f(x)$ and $g(x)$ in the same window for $-1 \leq x \leq 3$.

(c) Find an equation of a fourth degree polynomial $p(x)$ which passes through the point $(0, f(0))$, and satisfies the following derivative conditions.

$$p^{(k)}(0) = f^{(k)}(0) \text{ for } k = 1, 2, 3 \text{ and } 4.$$

Plot $f(x)$ and $p(x)$ in the same window for $-1 \leq x \leq 3$.

(d) Is one of $g(x)$ or $p(x)$ a better fit to $f(x)$? There are different ways to interpret "better fit." Can you define two different methods to measure the fit of one function to another?

Chapter 8

Inverse Trigonometric and Hyperbolic Functions

New Maple Commands for Chapter 8

arcsin(x);	function inverse to sin(x)
arctan(x);	function inverse to tan(x)
D(f)(g(x));	f'(g(x)), which is df/dx evaluated at x = g(x)

Introduction

The concept of an inverse function was explored in Chapter 7. Here in Chapter 8 we will study the calculus of inverse functions, and of the inverse trigonometric and inverse hyperbolic functions in particular. The fundamental relationship between the derivatives of inverse functions is that of reciprocals. We make this clear by examples, after we have discussed inverses for the trigonometric and hyperbolic functions.

Examples

Example 1

Obtain the inverse of the function sin(x).

Recalling the algorithm for inverse functions from Chapter 7, we set $y = \sin(x)$, solve for x, then switch the letters.

• q := y = sin(x);

$$q := y = \sin(x)$$

• q1 := solve(q, x);

$$q1 := \arcsin(y)$$

The function inverse to sin(x) has the name arcsin(x), and is built into Maple. The values of arcsin(x) would have to be tabulated from values of sin(x) if there were not other algorithms for computing it built into Maple. Perhaps a graph would be enlightening.

• plot(arcsin(x), x = -5..5);

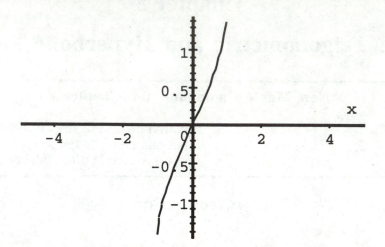

It would appear that in spite of our x = -5..5, the graph of arcsin(x) has a domain somewhere in the vicinity of [-1, 1] and a range of about [-1.6, 1.6]. If we reflect the graph of arcsin(x) across the line y = x, we will see the graph of sin(x) with which we are more familiar. From the graph of sin(x), we might extract the appropriate domain and range for arcsin(x).

• plot([arcsin(x), x, x = -1..1]);

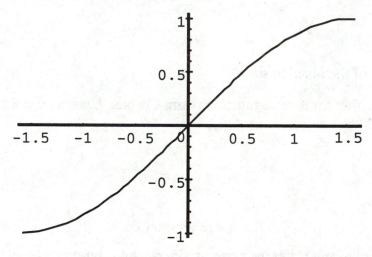

This is a monotone portion of the graph of sin(x) having domain [-π/2, π/2] and range [-1, 1]. Since the domain and range for inverse functions are the reverse of what they are for the original function, we have deduced that the domain of arcsin(x) is [-1, 1] and the range is [-π/2, π/2].

The reader is encouraged to repeat this study for the remaining five inverse trigonometric functions. It is especially useful to see what domain Maple chooses for these functions since all texts are not in universal agreement as to these domains.

Example 2

Find the inverse of the hyperbolic function sinh(x).

- q := y = sinh(x);

$$q := y = \sinh(x)$$

- q1 := solve(q, x);

$$q1 := \operatorname{arcsinh}(y)$$

The pattern is consistent. The function inverse to sinh(x) is arcsinh(x). We can determine the domain and range for this inverse much the same way as we did in Example 1 for arcsin(x). We leave for the reader the discovery that both the domain and range of arcsinh(x) are all the real numbers. Instead, we will show how to obtain an alternative representation of the function arcsinh(x). Begin with the exponential definition of sinh(x).

- q := y = simplify(convert(sinh(x), exp));

$$q := y = \frac{1}{2}e^x - \frac{1}{2}e^{(-x)}$$

Multiply both sides of equation q by e^x to eliminate the e^-x term.

- q1 := expand(exp(x)*q);

$$q1 := e^x y = \frac{1}{2}\left(e^x\right)^2 - \frac{1}{2}$$

Next, since this equation is quadratic in e^x,

- q2 := solve(q1, exp(x));

$$q2 := y - \sqrt{y^2 + 1}, y + \sqrt{y^2 + 1}$$

There are two branches, but e^x is monotone increasing for $x \geq 0$. This selects the second solution in q2 as an equivalent to e^x. Taking natural logarithms of both sides, and switching the letters x and y, leads to the same result as

- convert(arcsinh(x), ln);

$$\ln\left(x + \sqrt{x^2 + 1}\right)$$

We turn our attention now to the relationship between derivatives of inverse functions.

Example 3

Show that if f(x) and g(x) are inverse functions, their derivatives are related via the relationship g'(x) = 1/f'(g(x)).

Start with the recollection that f(x) and g(x) are related as inverse functions. Thus, f(g(x)) = x. Then differentiate with respect to x via the chain rule.

- q := x = f(g(x));

$$q := x = f(g(x))$$

- q1 := diff(q, x);

$$q1 := 1 = D(f)(g(x))\left(\frac{\partial}{\partial x} g(x)\right)$$

- readlib(isolate):
  isolate(q1, diff(g(x),x));

$$\frac{\partial}{\partial x} g(x) = \frac{1}{D(f)(g(x))}$$

That there is a reciprocal relation is clear. Maple's notation in the denominator on the right stands for f'(g(x)). And the **isolate** command was used to create an equation expressing this fundamental formula.

Example 4

Use the formula derived in Example 3 to obtain the derivative of arcsin(x).

Since the derivative of sin(x) is known to be cos(x), we take f(x) as sin(x) and g(x) as arcsin(x). Then, g'(x) = 1/cos(arcsin(x)).

- q := 1/cos(arcsin(x));

$$q := \frac{1}{\sqrt{1 - x^2}}$$

- diff(arcsin(x), x);

$$\frac{1}{\sqrt{1 - x^2}}$$

Example 5

Obtain the derivatives of the function sinh(x) and the inverse, arcsinh(x).

The derivative of sinh(x) is done the *eeezy* way.

- q := convert(sinh(x), exp);

$$q := \frac{1}{2}e^x - \frac{1}{2}\frac{1}{e^x}$$

- q1 := diff(q, x);

$$q1 := \frac{1}{2}e^x + \frac{1}{2}\frac{1}{e^x}$$

- q2 := convert(simplify(q1), trig);

$$q2 := \cosh(x)$$

- diff(sinh(x), x);

$$\cosh(x)$$

The reciprocal relation for the derivative of arcsinh(x) is implemented by taking f(x) as sinh(x) and g(x) as arcsinh(x).

- q3 := 1/cosh(arcsinh(x));

$$q3 := \frac{1}{\sqrt{x^2 + 1}}$$

- diff(arcsinh(x), x);

$$\frac{1}{\sqrt{x^2 + 1}}$$

Example 6

Find the maximum of arctan(2x) - arctan(x) on the domain x ≥ 0.

- q := arctan(2*x) - arctan(x);

$$q := \arctan(2\,x) - \arctan(x)$$

- plot(q, x = 0..10);

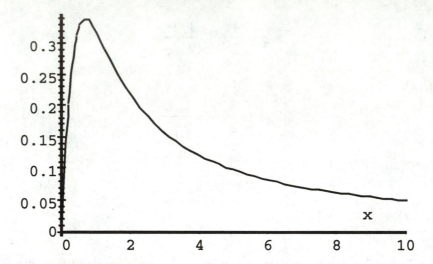

- q1 := diff(q, x);

$$q1 := 2\,\frac{1}{1+4\,x^2} - \frac{1}{x^2+1}$$

- q2 := solve(q1, x);

$$q2 := -\frac{1}{2}\sqrt{2}, \frac{1}{2}\sqrt{2}$$

Since the working domain is x ≥ 0, we choose the positive root and compute

- q3 := subs(x = 1/sqrt(2), q);

$$q3 := \arctan(\sqrt{2}) - \arctan\left(\frac{1}{2}\sqrt{2}\right)$$

- simplify(q3);

$$2\,\arctan(\sqrt{2}) - \frac{1}{2}\,\pi$$

- evalf(q3);

$$.3398369095$$

Exercises

1. (a) Plot sin x for -2 ≤ x ≤ 4. Identify three different intervals on the x-axis on which there is an inverse function of sin x. Make each interval as large as possible.

    (b) Solving the equation x = sin(y) for y and plotting should give the graph of arcsin(x). Execute the following commands and explain if this gives the graph of the function arcsin(x).

    with (plots):
    implicitplot (x = sin(y), x = -Pi..Pi, y = 0..10);

    (c) Plot arcsin(x) and compare with your plot in part (b).

2. Repeat Exercise 1, parts (a), (b), and (c) for tan x.

3. Plot the graph of $f(x) = \arcsin(x)\sqrt{1 - x^2}$, and find the maximum value of f(x).

4. Define $f(x) = \arcsin(\frac{x}{3})$, and g(x) = ln(x).

    (a) In the same window, plot f(x) and g(x) for 0.2 ≤ x ≤ 15.

    (b) Do f(x) and g(x) intersect?

    (c) Replot f(x) and g(x) for 1.9 ≤ x ≤ 2.2.

    (d) Find the minimum value of f(x) - g(x).

    (e) Does your result in part (d) tell how close the curves approach each other?

5. Derive the differentiation formula for arcsin(x) by writing y = arcsin(x) as x = sin(y), then differentiating and using the commands **solve** and **simplify**.

6. Repeat Exercise 5 for arctan(x).

7. Show that $\arctan(\frac{x-1}{x+1}) = \arctan(x) - \frac{\pi}{4}$. (First, find the derivative of $\arctan(\frac{x-1}{x+1})$.)

8. Plot each function and decide where an inverse function exists.

    (a) f(x) = sinh(x)   (c) f(x) = arcsinh(x)

    (b) f(x) = cosh(x)   (d) f(x) = arccosh(x).

9.   Derive the differentiation formula for arccosh(x) .

10. An arch is defined as the portion of the curve $y = 6 - \cosh(\frac{x-2}{6})$ which is above the x-axis.

   (**a**) Plot the curve on the interval for which y is nonnegative.

   (**b**) Find the maximum height of the arch.

   (**c**) Find the area of the region below the arch.

11. Find all points of intersection of $f(x) = \cosh x$ and $g(x) = 3 + \ln x$.

12. Find the area of the region that is between the graphs of $y = \cosh x$ and $y = 3 + \ln x$.

13. Find the volume of the solid formed when the region between $y = \sinh x$, $y = 0$, and $x = 1$ is revolved about the x-axis.

Projects

1.   A right circular cone has slant height 10. Find the dimensions that will maximize

   (**a**) the volume   (**b**) the ratio of volume to surface area.

2.   Define $f(x) = \cos(x)$ and plot f(x) for $x \in [-\frac{1}{2}..4]$. Find f(0.2), f(1.0), f(1.5), and f(3.5). Locate the corresponding points on your plot. Here you are asked to plot 41 points on g(x), the inverse of f(x). Note, for example, that f(1.0) = 0.54, so g(0.54) = 1.0. The point (1.0, 0.54) appears on the graph of f(x), while the graph of g(x) contains the point (0.54, 1.0). Hence, a graph of g(x) can be constructed from points (x, f(x)) if the order of the coordinates is reversed to (f(x), x). Try plotting g(x) via the command plot([seq([sin(-1/2+k/10),-1/2+k/10],k=0..40)]).

   Why does g(x) fail to be a function? How should we restrict the domain of f(x) in order to obtain an inverse function? On this smaller domain, plot f(x) parametrically, along with a parametric plot of the function that has its coordinates reversed. Compare your plot to the graph of the function arccos(x) used by Maple. What is the domain of the inverse function? The range?

3.   Repeat Exercise 2 with $f(x) = \sin(x)$. Start with $x \in [-2..3]$.

4.   Define $f(x) = \arctan(\frac{x+2}{x}) + \arctan(x + 1)$.

(a) Find decimal values of f(0.5), f(1), and f(10).

(b) Plot f(x) for $0 \le x \le 15$.

(c) Prove that f(x) is a constant-valued function.

(d) Does an inverse function exist for f(x)?

5.   Define $f(x) = \dfrac{x}{1 + x^2}$ and g(x) = arctan(x).

(a) Plot f(x) and g(x) for $0 \le x \le 5$; for $0 \le x \le 0.2$.

(b) Compare decimal values of f(0.1) and g(0.1); f(0.02) and g(0.02).

(c) Prove that $f(x) \le g(x)$ for all $x \ge 0$. (Hint: Apply the Mean Value Theorem to the arctan function on the interval [0, x].)

6.   (a) Plot the graph of the function arccosh(x).

(b) Obtain an implicitplot of the equation y = cosh(x).

(c) Are your plots the same?

7.   Walls A and B are perpendicular, and a camera moves along wall B and looks through a window on wall A. The window is 5 feet wide and the end of the window is 4 feet from wall B.

(a) Define a function that gives the angle at the camera subtended by the window, as a function of the camera distance from wall A.

(b) Plot the graph of your angular function.

(c) Find the maximum value of this angle.

8.   Make a nine-frame movie of the angle in Problem 7 as the camera moves from the point that is 10 feet from wall A to the point that is 1 foot from wall A. Make a second movie that consists of 27 frames, and compare the motion in your two animations.

Chapter 9

Integration - Exact and Approximate

New Maple Commands for Chapter 9

convert(f, parfrac, x);	partial fraction decomposition of rational function f(x)
evalf(Int(f,x=a..b));	numeric evaluation of definite integral
normal(expression);	add fractions via common denominator
simplify(expression,symbolic);	forces Maple to use transformation $\sqrt{x^2} \to x$
with(student);	loads student package
changevar(eqn,Int,t);	makes change of variables defined by eqn in inert integral Int, resulting in integral with new variable t
intparts(Int,u);	gives u v - $\int$v du for the inert integral Int = $\int$u dv
simpson(f,x=a..b,n);	Simpson's Rule approximation with n panels
trapezoid(f,x=a..b,n);	Trapezoidal Rule approximation with n panels

---

Introduction

The Fundamental Theorem of Calculus allows us to evaluate definite integrals by finding antiderivatives. The techniques for finding these antiderivatives constitute an art, and for some, it is a religion. In this chapter we explore some of the strategies for "evaluating" integrals. For indefinite integrals, evaluation means finding an antiderivative. For definite integrals, we either find an appropriate antiderivative and produce a value for the integral, or we approximate the integral numerically.

There are many methods for numerically evaluating definite integrals. Maple has a sophisticated numeric integrator built in, and it also allows students to explore the Trapezpoidal Rule and Simpson's Rule found in the student package. Maple's built-in numeric integrator is an adaptive quadrature that puts additional function evaluations into subintervals where needed. Hence, accuracy and efficiency of the built-in integrator cannot be matched by naive applications of even Simpson's Rule.

The issue of "methods of integration" has been vigorously debated during the transition from pencil-and-paper-based traditional calculus courses to courses in which software like Maple is available. The concepts underlying changing the form of an integral by a substitution, integrating by parts, and decomposing a rational function by partial fractions are all important milestones in understanding and using mathematics. However, no one makes a living from the skills of evaluating integrals by any of these techniques, and virtuosity in them is misplaced energy. We take the point of view that familiarity is essential, but few students ever need to evaluate "by hand" integrals that appear in tables of integrals. We offer the reader examples of explorations by which students can gain the required understanding of the techniques of integration.

Examples

Example 1

Find the volume of the solid formed when the region in the first quadrant, bounded by the curves x = 0, y = arcsin(x), and y = 1 - ln(x), is rotated about the x-axis.

We will first want to look at a graph of the region.

- f := arcsin(x);
  g := 1 - ln(x);
  plot({f, g}, x = 0..1);

$$f := \arcsin(x)$$

$$g := 1 - \ln(x)$$

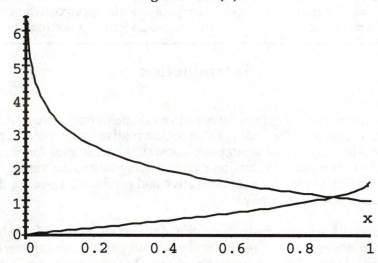

Next, we want the point of intersection of the two curves. No formula exists to express the solution to the transcendental equation f = g, so we obtain a numeric solution via **fsolve**.

- a := fsolve(f = g, x);

$$a := .8957436147$$

The required volume is given by an integral of the form $\pi \int [(yupper)^2 - (ylower)^2]\, dx$.

- vol := Pi * Int(g^2 - f^2, x = 0..a);

$$vol := \pi \int_0^{.8957436147} (1 - \ln(x))^2 - \arcsin(x)^2 \, dx$$

- evalf(vol);

$$14.40316195$$

Maple's built-in numeric integrator is triggered when the argument of **evalf** is an unevaluated integral. We created our unevaluated integral by the inert integration operator **Int**. Had we attempted the integral using **int**, Maple would have returned it unevaluated since the antiderivative of the square of the arcsin is not an elementary function.

- value(vol);

$$\pi \int_0^{.8957436147} (1 - \ln(x))^2 - \arcsin(x)^2 \, dx$$

When it is known beforehand that the antiderivative is beyond Maple (or mathematics) to express in terms of the simple functions, it is faster to use **Int**, which is unevaluated, and **evalf**, which then immediately triggers a numeric integration without the intervention of a search for an unobtainable antiderivative.

Example 2

Approximate the following integral by using Maple's built-in integrator, the Trapezoidal Rule and Simpson's Rule, both in the student package.

- q := Int(arcsin(x)^2, x = 1/2..1);

$$q := \int_{\frac{1}{2}}^1 \arcsin(x)^2 \, dx$$

- with(student):
- q1 := evalf(q);

$$q1 := .4234235793$$

- qq2 := trapezoid(arcsin(x)^2, x = 1/2..1, 10);

$$qq2 := \frac{1}{144}\pi^2 + \frac{1}{20}\left(\sum_{i=1}^{9} \arcsin\left(\frac{1}{2} + \frac{1}{20}i\right)^2\right)$$

- q2 := evalf(value(qq2));

$$q2 := .4330866500$$

- qq3 := simpson(arcsin(x)^2, x = 1/2..1, 10);

$$qq3 := \frac{1}{216}\pi^2 + \frac{1}{15}\left(\sum_{i=1}^{5} \arcsin\left(\frac{9}{20} + \frac{1}{10}i\right)^2\right) + \frac{1}{30}\left(\sum_{i=1}^{4} \arcsin\left(\frac{1}{2} + \frac{1}{10}i\right)^2\right)$$

- q3 := evalf(value(qq3));

$$q3 := .4274530821$$

It appears that Simpson's Rule is more accurate.

- abs(q1 - q2), abs(q1 - q3);

$$.0096630707, .0040295028$$

In fact, Simpson's Rule is nearly twice as accurate in this calculation, but with just 10 panels even Simpson's Rule is accurate to only three places. Since a formula for the error made by Simpson's Rule is available, we can estimate how much of an error this calculation might make when evaluating with n = 10.

Example 3

Estimate the error made by Simpson's Rule (with n = 10) when it is used to approximate the integral

- f := sin(x^2);
  Int(f, x = 0..1);

$$f := \sin(x^2)$$

$$\int_0^1 \sin(x^2)\, dx$$

The error is known to obey the formula Error = $(b-a)^5 f^{(4)}(c)/(180\, n^4)$, where c is some point in the interval [0, 1], and $f^{(4)}$ is the fourth derivative of f(x). Now the point c is generally not obtainable, so the error estimate is used in a "worst case" sense. The worst the error can be is what we would compute if we took the maximum of the fourth derivative over the interval [0, 1]. Thus, |Error| ≤ $(b-a)^5 \text{Max}(|f^{(4)}(x)|)/(180\, n^4)$. So, take b - a as 1 - 0 = 1, n = 10, and $f^{(4)}(x)$ as

- f4 := diff(f, x$4);

$$f\!4 := 16 \sin(x^2)\, x^4 - 48 \cos(x^2)\, x^2 - 12 \sin(x^2)$$

- plot(abs(f4), x = 0..1);

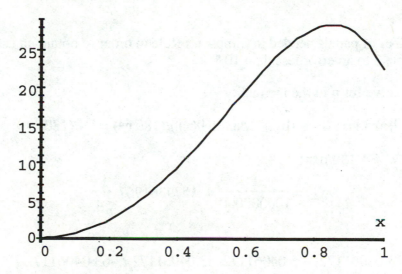

The maximum of f4 appears to occur at an interior point in the interval [0, 1]. Suppose we estimate this maximum by the methods of Chapter 4.

- q := fsolve(diff(f4,x) = 0, x, 0..1);

$$q := .8520766663$$

- F4 := abs(evalf(subs(x = q, f4)));

$$F\!4 := 28.42851540$$

- maxerror := F4/180/10^4;

$$maxerror := .00001579361967$$

Maple's built-in integrator obtains

- q1 := evalf(Int(f, x = 0..1));

$$q1 := .3102683017$$

and Simpson's Rule, for n = 10, obtains

- q2 := evalf(value(simpson(f, x = 0..1, 10)));

$$q2 := .3102602344$$

so the actual error made by Simpson's Rule is

- abs(q1 - q2);

$$.80673 \ 10^{-5}$$

which is about half the estimate obtained in maxerror.

Example 4

Estimate the number of panels needed in Simpson's Rule in order to obtain an estimate of the integral in Example 3 to an error less than 10^{-8}.

Here, we need to solve for n in the inequality

$$|\text{Error}| = 10^{-8} \le (b-a)^5 \ \text{Max}(|f^{(4)}(x)|)/(180 \ n^4) = F4/(180 \ n^4)$$

- Q := 10^(-8) = F4/180/n^4;

$$Q := \frac{1}{100000000} = .1579361967 \ \frac{1}{n^4}$$

- solve(Q, n);

$$63.04061172, -63.04061172, 63.04061172 \ I, -63.04061172 \ I$$

The real solution of 63.04 tells us that when using Simpson's Rule to estimate the integral in Example 3 with an error less than 10^{-8} we must use n = 64 to be sure. It might be that a smaller n will give us the required accuracy, but the error estimate we used requires n = 64 for the assurance we seek.

- for k from 50 by 2 to 56 do
      k, evalf(q1 - value(simpson(f, x = 0..1, k)));
  od;

$$50, .12620 \ 10^{-7}$$

$$52, .10965 \ 10^{-7}$$

$$54, .956 \ 10^{-8}$$

$$56, .835 \ 10^{-8}$$

In fact, n = 54 suffices to give, by Simpson's Rule, an error of less than 10^{-8}.

Example 5

Use an appropriate trigonometric substitution to evaluate the following indefinite integral.

• q := Int(x/sqrt(3 - 8*x^2), x);

$$q := \int \frac{x}{\sqrt{3 - 8 \, x^2}} \, dx$$

The trigonometric substitution we try is defined by the following equation.

• qq := 8*x^2 = 3*sin(t)^2;

$$qq := 8 \, x^2 = 3 \, \sin(t)^2$$

The transformed integral results from an application of Maple's **changevar** command, found in the student package.

• q1 := changevar(qq, q, t);

$$q1 := \int \frac{3}{8} \frac{\sin(t) \cos(t)}{\sqrt{3 - 3 \, \sin(t)^2}} \, dt$$

To simplify the integrand, we use the *symbolic* option so that the transformation $\sqrt{(x^2)} \to x$ is made.

• q2 := simplify(q1, symbolic);

$$q2 := \frac{1}{8} \sqrt{3} \int \sin(t) \, dt$$

The value of this integral is

• q3 := value(q2);

$$q3 := -\frac{1}{8}\sqrt{3}\,\cos(t)$$

We next return from t to x by solving the transformation equation for t = t(x) and substituting t = t(x) in q3.

• T := solve(qq, t);

$$T := -\arcsin\left(\frac{2}{3}\sqrt{6}\,x\right),\ \arcsin\left(\frac{2}{3}\sqrt{6}\,x\right)$$

• q4 := subs(t = T[2], q3);

$$q4 := -\frac{1}{8}\sqrt{3}\,\cos\left(\arcsin\left(\frac{2}{3}\sqrt{6}\,x\right)\right)$$

• q5 := simplify(q4);

$$q5 := -\frac{1}{24}\sqrt{3}\,\sqrt{9 - 24\,x^2}$$

A direct call to Maple's integrator yields

• value(q);

$$q5 := -\frac{1}{8}\sqrt{3 - 8\,x^2}$$

which differs from q5 only in how √3 was treated.

Example 6

Explore the technique of integration by parts by using Maple's **intparts** command (in the student package) to study the effect of different choices of u in integrands of the form ∫u dv. In each case, state clearly what u, du, dv, and v are. Then conclude which choice of u led to the "simplest" new integral of the form ∫v du.

• q := Int(x*exp(x), x);

$$q := \int x\,e^x\,dx$$

• q1 := intparts(q, x);

$$q1 := x\,e^x - \int e^x\,dx$$

For q1 we have u = x, du = dx, dv = e^x dx, and v = e^x. Hence, v du is e^x dx.

- q2 := intparts(q, exp(x));

$$q2 := \frac{1}{2}e^x x^2 - \int \frac{1}{2}e^x x^2 \, dx$$

For q2 we have u = e^x, du = e^x dx, dv = x dx, and v = x^2/2. Hence, v du is x^2/2 e^x.

- q3 := intparts(q, x*exp(x));

$$q3 := e^x x^2 - \int (e^x + x e^x) x \, dx$$

For q3 we have u = x e^x, du = (x +1) e^x dx, dv = dx, and v = x. Hence, v du is x(x + 1)e^x.

- q4 := intparts(q, 1);

$$q4 := x e^x - e^x - \int 0 \, dx$$

For q4 we have u = 1, du = 0, dv = x e^x dx, and v = $\int$xe^x dx, the integral we started with. So, the simplest new integral to perform is found in q1 with a choice of u = x.

Example 7

Decompose the following rational function by the algebraic technique of partial fractions.

- f := (x+3)/(x^3 - 2*x^2 - 8*x);

$$f := \frac{x+3}{x^3 - 2x^2 - 8x}$$

Maple has an algorithm for generating the partial fraction decomposition.

- convert(f, parfrac, x);

$$-\frac{3}{8}\frac{1}{x} + \frac{1}{12}\frac{1}{x+2} + \frac{7}{24}\frac{1}{x-4}$$

Example 8

Obtain the partial fraction decomposition of Example 7 by directing Maple to implement the steps of some form of the algorithm that you would learn in a traditional calculus course.

Begin by factoring the denominator of f.

- factor(denom(f));

$$x(x+2)(x-4)$$

These factors tell us the form of the fractions to expect in the decomposition.

- f1 := A/x;
  f2 := B/(x + 2);
  f3 := C/(x - 4);

$$f1 := \frac{A}{x}$$

$$f2 := \frac{B}{x+2}$$

$$f3 := \frac{C}{x-4}$$

Next, add these fractions.

- q := f1 + f2 + f3;

$$q := \frac{A}{x} + \frac{B}{x+2} + \frac{C}{x-4}$$

To make Maple *really* add these fractions, use the **normal** command.

- q1 := normal(q);

$$q1 := \frac{A x^2 - 2 A x - 8 A + B x^2 - 4 B x + C x^2 + 2 C x}{x(x+2)(x-4)}$$

The numerators of f and q1 must match exactly, term for term. Equating coefficients of like powers of x in these two numerators gives us three equations in the three unknowns A, B, and C.

- q2 := solve(identity(numer(q1) = numer(f), x), {A, B, C});

$$q2 := \left\{ B = \frac{1}{12}, C = \frac{7}{24}, A = \frac{-3}{8} \right\}$$

The purist who insists on obtaining and solving the governing equations can use the following Maple instructions.

- for k from 0 to 2 do   e.k := coeff(numer(q1), x, k) = coeff(numer(f), x, k);   od;

$$e0 := -8\,A = 3$$

$$e1 := -4\,B - 2\,A + 2\,C = 1$$

$$e2 := A + B + C = 0$$

- solve({e0, e1, e2}, {A, B, C});

$$\left\{ B = \frac{1}{12}, C = \frac{7}{24}, A = \frac{-3}{8} \right\}$$

However the values of A, B, and C were computed, the actual decomposition can be obtained by

- q3 := subs(q2, q);

$$q3 := -\frac{3}{8}\frac{1}{x} + \frac{1}{12}\frac{1}{x+2} + \frac{7}{24}\frac{1}{x-4}$$

Example 9

Obtain $\int f\,dx$ for the function f in Example 7.

The context of partial fraction decomposition is usually integration since a complicated rational function is thereby expressed as a sum of fractions with simpler denominators. In practice, however, few people ever have the need to perform such integrals. Partial fraction decomposition is used much more frequently in conjunction with Laplace transforms, especially in the design of feedback control systems. For those who still feel that the context for partial fraction decomposition is antidifferentiation, we find the required definite integral from expression q3 given in Example 8.

- int(f, x);

$$-\frac{3}{8}\ln(x) + \frac{1}{12}\ln(x+2) + \frac{7}{24}\ln(x-4)$$

- int(q3, x);

$$-\frac{3}{8}\ln(x) + \frac{1}{12}\ln(x+2) + \frac{7}{24}\ln(x-4)$$

Exercises

1. Use substitution, that is, a change of variable, to evaluate each integral without using Maple. Then, to verify your answers, use the commands **changvar**, **Int** (not **int**), **value**, and **subs**.

    (a) $\int x^2 (x^3 + 5)^6 \, dx$ **(c)** $\int \sin x \cos^3 x \, dx$

    (b) $\int x^3 \sqrt{x^4 + 1} \, dx$ **(d)** $\int \tan^3 x \sec^2 x \, dx$.

2. Define $f(x) = \ln(x)$ and $g(x) = 3 + 2x - x^2$.

    (a) In the same window, plot $f(x)$ and $g(x)$ for $\frac{1}{2} \le x \le 4$.

    (b) Find the area of the region in the first quadrant which is below both $f(x)$ and $g(x)$.

    (c) Find the volume of the solid that is formed by revolving the region in part **(b)** about the x-axis.

3. Define $f(x) = \dfrac{x}{x^2 - 1}$.

    (a) Use the command **convert**$(f(x), \text{parfrac}, x)$ to write $f(x)$ in terms of partial fractions.

    (b) Integrate your result from part **(a)**.

    (c) Check your answer by applying **diff** and **simplify**.

    (d) Use the commands **value** and **Int** to solve this problem directly.

4. Repeat the directions from Exercise 3 for $f(x) = \dfrac{3x^2 - 2x - 17}{x^3 - 2x^2 - x + 2}$.

5. Repeat the directions from Exercise 3 for $f(x) = \dfrac{x^3 + x^2 - 5x - 7}{x^4 - x^3 + 4x^2 - x + 3}$.

6. **(a)** Use Integration by Parts to derive the reduction formula

$$\int \sec^n x \, dx = \frac{1}{n-1} \sec^{n-2} x \tan x - \frac{n-2}{n-1} \int \sec^{n-2} x \, dx.$$

(**b**) Find a reduction formula for $\int x^m (\ln x)^n \, dx$.

7. Pick an appropriate trigonometric substitution and see if Maple can eliminate the radical:

(**a**) $\sqrt{25 - x^2}$ (**b**) $\sqrt{4 + 5x^2}$ (**c**) $\sqrt{5x^2 + 10x + 1}$.

8. (**a**) Find the arc length of the parabola $y = x^2$ from $x = 0$ to $x = 4$.

(**b**) Find the indefinite integral (the antiderivative) corresponding to your part (**a**) definite integral.

9. Define $f(x) = x^2 + 1$.

(**a**) Display the 10 inscribed approximating rectangles to the area under $f(x)$ from $x = 0$ to $x = 3$, using the command **leftbox**. (Don't forget 'with(student)'.)

(**b**) Approximate the area under $f(x)$ from $x = 0$ to $x = 3$ by summing the areas of the 10 inscribed rectangles using the command **sum**.

(**c**) Approximate the area under $f(x)$ from $x = 0$ to $x = 3$ by summing the areas of the 10 inscribed rectangles using the command **leftsum**.

10. Estimate the area under $f(x) = \sqrt{x^2 + 1}$ in the first quadrant, from $x = 0$ to $x = 3$, by dividing the interval $[0,3]$ into 20 equal subintervals, and using the Maple command

(**a**) **middlesum**   (**b**) **trapezoid**   (**c**) **simpson**.

Error bounds for these three approximating procedures using n subintervals are:

$$|E_M| \le \frac{f2}{24n^2}(b - a)^3, \quad \text{where } |f^{(2)}(x)| \le f2 \text{ on } [a,b],$$

$$|E_T| \le \frac{f2}{12n^2}(b - a)^3, \quad \text{where } |f^{(2)}(x)| \le f2 \text{ on } [a,b],$$

$$|E_S| \le \frac{f4}{180n^4}(b - a)^5, \quad \text{where } |f^{(4)}(x)| \le f4 \text{ on } [a,b].$$

(**d**) Find error bounds for each of your answers in parts (**a**), (**b**), and (**c**).

11. Use the First Fundamental Theorem of Calculus to find the exact first quadrant area under $f(x) = \sqrt{x^2 + 1}$, $0 \le x \le 3$. Compare your answer with each result from

Exercise 10, parts (**a**), (**b**) and (**c**). Do the errors agree with the error bounds from Exercise 10, part (**d**)?

12. Explain briefly why you cannot use the Trapezoidal Rule or Simpson's Rule to find an answer to $\int e^{x^3} dx$.

13. Define $f(x) = \sqrt{4 + x^3}$.

   (**a**) Use the Trapezoidal Rule to estimate $\int_0^{1.7} f(x)\ dx$, taking $n = 10$; $n = 50$. Find an error bound in each case.

   (**b**) Find the smallest integer n which guarantees that the estimate in part (**a**) has error $\le 0.5\ 10^{-5}$.

   (**c**) Use the Simpson Rule to estimate $\int_0^{1.7} f(x)\ dx$, taking $n = 10$; $n = 50$. Find an error bound in each case.

   (**d**) Find the smallest integer n which guarantees that the estimate in part (**c**) has error $\le 0.5\ 10^{-10}$.

14. Define $f(x) = \sqrt{1 - \sin x \cos x}$.

   (**a**) Use the Trapezoidal Rule to estimate $\int_0^{1.5} f(x)\ dx$, taking $n = 10$; $n = 25$. Find an error bound in each case.

   (**b**) Find the smallest integer n which guarantees that the estimate in part (**a**) has error $\le 0.5\ 10^{-5}$.

   (**c**) Use the Simpson Rule to estimate $\int_0^{1.5} f(x)\ dx$, taking $n = 10$; $n = 26$. Find an error bound in each case.

   (**d**) Find the smallest integer n which guarantees that the estimate in part (**c**) has error $\le 0.5\ 10^{-10}$.

15. Approximate π by using Simpson's Rule on $\int_0^1 \dfrac{1}{x^2 + 1}\ dx$. Estimate the error in your answer.

16. Can you find an approximation to $\int_0^1 \dfrac{\sin x}{\sqrt{x}}\, dx$? There is a serious complication here; do you recognize it? (The next chapter will deal with this kind of problem.)

17. Use Maple to derive the Simpson Rule for $\int_{x=a}^{x=a+2h} f(x)\, dx$ by completing the following outline:

   i)   Determine A, B, and C such that the parabola $y = A x^2 + B x + C$ passes through the points (a, y0), (a+h, y1), and (a+2h, y2).

   ii) With these values of A, B, and C, find $\int_{x=a}^{x=a+2h} (A x^2 + B x + C)\, dx$ and write this result in the form of Simpson's rule.

Projects

1. Let R be the region between the curves $y = 0$ and $y = e^{x^2}$, from $x = 0$ to $x = 1.5$. Estimate the area of R correct to two decimal places.

2. Let R be the region between the curves $y = 0$ and $y = \sqrt{4 + x^3}$, from $x = 0$ to $x = \sqrt[3]{5}$. Estimate the area of R correct to four decimal places.

3. **(a)** Write Maple instructions to approximate the area under $y = f(x)$ and above $y = 0$, from $x = a$ to $x = b$, by summing the areas of n trapezoids having equal bases on [a, b].

   (b) Test your code for $y = x^2 + 1$ with $n = 10$ on [a, b] = [1, 3]. Compare your answer with the result of using the **trapezoid** command.

4. **(a)** Write Maple instructions to approximate the area under $y = f(x)$ and above $y = 0$ from

   $x = a$ to $x = b$, by repeated ($\frac{n}{2}$ times) use of the Simpson formula

   $A_i = \dfrac{h}{3}(y_i + 4y_{i+1} + y_{i+2}).$

   (b) Test your code for $y = e^{x^2}$ with $n = 10$ on [a, b] = [0, 1.5]. Compare your answer with the result of using the **simpson** command.

5. For the problem given, find a Trapezoidal approximation using n = 20 and a Simpson approximation using n = 10. Find a bound on the error associated with each approximation. Then find the smallest integer n for which the Trapezoid error is less than $0.5 \ 10^{-5}$, and also the smallest integer n for which the Simpson error is less than $0.5 \ 10^{-8}$.

    (a) $\int_0^{1.5} \sqrt{1 - \sin^2 x \cos x} \ dx$ (c) $\int_0^1 \dfrac{1}{\sqrt{x^3 + 0.2}} \ dx$

    (b) $\int_0^{1.5} \sqrt{1 - \sin x \cos^2 x} \ dx$ (d) $\int_0^2 (x + 1 - \sin x)^{0.3} \ dx$.

6. Define $f(x) = 6 x - x^2 + 3$.

    (a) Display the eight inscribed rectangles having equal bases on the interval [0, 4]. In a separate window display the eight circumscribed rectangles.

    (b) Repeat part (a) using nine inscribed rectangles. Because the high point on the graph of f(x) is not assumed at an endpoint of a base of any of the rectangles, you will need to spend some effort deciding how many of the rectangles should be generated by each of the **leftbox** and **rightbox** commands.

    (c) Now, we will ask you to increase the number of inscribed rectangles to at least 50, and then to increase the number of circumscribed rectangles to the same number. Instead of the '__box' commands used to display the rectangles, use the **leftsum** and **rightsum** commands to compute the total area of the inscribed rectangles. Do the same for the circumscribed rectangles.

    To make it easier to decide how to distribute the rectangles in the **leftsum** and **rightsum** commands, you should select the total number of rectangles so that the endpoint of a base falls on the abscissa of the high point. This will allow you to avoid the complications faced when using nine rectangles as in part (b).

    (d) Again increase the number of rectangles, to at least twice as many as you used in part (c), and find the total of the areas. What is happening to this total area?

7. Use the command **intparts** in the student package to step through the following integration problems. The first problem is solved for you. (Output not shown.) For each return from the **intparts** command, identify and label u, du, v, and dv.

    (a) $\int x^2 \cos(x) \ dx$
    - q1 := intparts(Int(x^2 * cos x, x), x^2);
    - q2 := intparts(q1, x);
    - q3 := value(q2);

(b) $\int x^2 \ln(x)\, dx$

(c) $\int x \sin(x)\, dx$

(d) $\int x^2 e^x\, dx$

(e) $\int e^x \cos(x)\, dx$.

8. Consider a right angle. A circle of unit radius is constructed such that its center lies on the bisector of the right angle and one-half of its area lies within the right angle. What is the distance from the center of the circle to the vertex of the right angle?

-P. B. Silverman (The Bent)

9. Make a movie that shows unit circles (as in Problem 8), whose centers move from the vertex of the right angle along the bisector. Choose the frame which shows that circle having one-half of its area within the right angle.

10. What is the ratio of the area of an ellipse (with major and minor axes a and b, respectively) to the area of the largest rectangle that can be incribed in it?

-The Crucible (The Bent)

11. Write Maple instructions to perform numerical integration in a fashion similar to Simpson's Rule, but instead of a parabola through three points, use a cubic through four points.

12. To introduce you to the idea of an adaptive procedure, we ask that you execute each of the two following plot commands, noting that Maple is being directed to plot 10 points in each case.

• plot (sin(x), x = 0..1, style = POINT, symbol = DIAMOND, numpoints = 10);
• plot (sin(x), x = 0.5..1.5, style = POINT, symbol = DIAMOND, numpoints = 10);

You should have more points appearing on the second plot, the reason being that Maple detected that the second graph had greater curvature, and so adapted the process of plotting to include additional points in the region of greater curvature.

In this project you are asked to create an adaptive Simpson procedure, that is, one that computes a Simpson approximation to $\int_a^b f(x)\, dx$, checks the maximum error for a specified tolerance, and, if needed, recomputes a Simpson approximation using a larger number of subintervals. Experience with recursive programming will be very helpful in understanding the following outline of this project.

Adaptive Simpson Algorithm to estimate $\int_a^b f(x)\,dx$ with an error $\leq \varepsilon$:

s1 := simpson value for 2 subintervals on [a, b]
s2 := simpson value for 4 subintervals on [a, b]
if $|s1 - s2| < 15\varepsilon$ then do
   's2 is a satisfactory answer'
else
   apply procedure to $[a, \frac{a+b}{2}] \to$ s21

   apply procedure to $[\frac{a+b}{2}, b] \to$ s22

   s := s21 + s22
od

Test your algorithm on $\int_1^3 \sqrt{x^3 + 2}\,dx$ with $\varepsilon = 0.0002$.

Chapter 10

Improper Integrals

New Maple Commands for Chapter 10

additionally(s<1);	adds additional assumption on s after an **assume**
int(f,x=a..b,CauchyPrincipalValue);	compute Cauchy Principal Value of improper integral

---

Introduction

Riemann integration, the integration studied in most introductory calculus courses, is defined for bounded functions on a bounded interval. It is an extension of the theory of Riemann integration to allow unbounded regions or unbounded integrands in the context of Riemann integration.

Definite integrals with unbounded integrands or over unbounded intervals are called improper integrals, and in Chapter 10 we study these extensions to the theory of integration. Since many functions in applied mathematics are defined by such improper integrals, it is useful to master the key concepts at this time.

Some texts distinguish between improper integrals of the First, Second, and Third Kinds. Some texts do not make any distinctions at all. But if your text mentions improper integrals of the Second Kind, they are the ones with the integrand becoming infinite at a point in the interval of integration and improper integrals of the First Kind are the ones where the interval of integration is infinite. If both things happen to the integral, it is called an improper integral of the Third Kind. So if you understand the first and second of these cases, you actually understand the third. And there is nothing like a few good illustrative examples to make any topic clear.

Examples

Example 1

The function $f = 1/(1 + x^2)$ is shaped like a bell with an infinite "flare." Find the area under this flare if x is in the interval $[1, \infty]$.

- f := 1/(1 + x^2);

$$f := \frac{1}{1 + x^2}$$

- plot(f, x = -5..5);

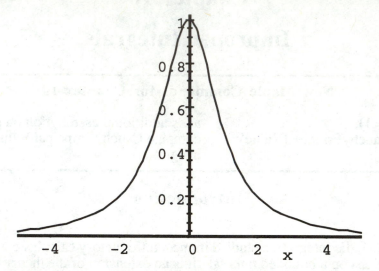

Since Riemann integration is properly defined for finite intervals, integrate from $x = 1$ to $x = z$, where $z > 1$ is some finite value. Thus, the integral we next write can be considered as prototypical of what such unbounded intervals will require.

- q := Int(f, x = 1..z);

$$q := \int_{1}^{z} \frac{1}{1 + x^2}\, dx$$

- q1 := value(q);

$$q1 := \arctan(z) - \frac{1}{4}\pi$$

Infinite intervals might be improper for Riemann integration, but they are certainly within the purview of limits. The limit of q1 as $z \to \infty$ will *define* the improper integral.

- q2 := limit(q1, z = infinity);

$$q2 := \frac{1}{4}\pi$$

Maple could have obtained this result directly.

- int(f, x = 1..infinity);

$$\frac{1}{4}\pi$$

Example 2

Obtain the value of the following improper integral.

- q3 := Int(t*exp(-s*t), t = 0..infinity);

$$q3 := \int_0^\infty t\, e^{(-s\,t)}\, dt$$

Moved by Maple's success with the improper integral in Example 1, we try

- q4 := value(q3);

$$q4 := \lim_{t \to \infty-} \frac{-s\, t\, e^{(-s\,t)} - e^{(-s\,t)}}{s^2} + \frac{1}{s^2}$$

Well, it seems Maple integrated to a finite upper limit (t) and correctly indicated the need for a limit as t → ∞. Maple could not compute this limit because its value depends on s. If s is positive, then the limit of e-st will be zero. If s is negative, the limit of e-st is infinite. Hence, we need to tell Maple that we would like s to be considered positive.

- assume(s>0);
- value(q4);

$$\frac{1}{s\!\sim^2}$$

As soon as Maple knew that s was positive, it completed the computation of the improper integral in q3. The tilde (~) attached to the s reminds us that the value of the integral is 1/s² only if s > 0.

This improper integral of the First Kind is an example of a Laplace transform, so important in the applications. And it illustrates how an improper integral can define a function, in this case 1/s², by "integrating out t" and leaving just the parameter s behind.

The last thing to note is how to remove the assumption on s.

- s := 's';

$$s := s$$

Example 3

Evaluate the following improper integral of the First Kind that cannot be computed exactly by any known antiderivative.

- f := cos(x)/(1 + x^5);

$$f := \frac{\cos(x)}{1 + x^5}$$

- q5 := Int(f, x = 0..infinity);

$$q5 := \int_0^\infty \frac{\cos(x)}{1 + x^5} dx$$

A graph of the integrand of the integral in q5 shows that the oscillations from cos(x) in the numerator are rapidly damped out by the $1 + x^5$ in the denominator. Hence, integrating to a modest finite upper limit should yield an accurate approximation to the improper integral.

- plot(f, x = 0..Pi);

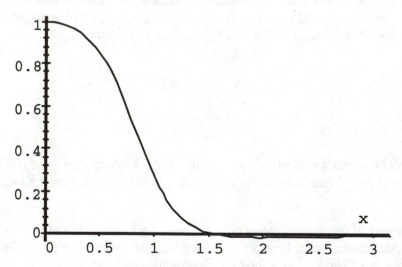

- evalf(Int(f, x = 0..10));

.8022814977

We leave the continuance of this computation to the reader and address, instead, the issue of convergence. Since the integrand changes sign infinitely often, the convergence will be conditional, unless we can show absolute convergence by showing $\int_R |f| \, dx$ is dominated by a convergent integral. But since $|\cos(x)| \leq 1$, we study the convergence of

• q6 := Int(1/(1 + x^5), x = 0..infinity);

$$q6 := \int_0^\infty \frac{1}{1+x^5}\,dx$$

whose value is

• value(q6);

$$\frac{1}{25}\pi\sqrt{5}\sqrt{2}\sqrt{5+\sqrt{5}}$$

Indeed, the dominating integral converges, and therefore so does the integral in q5.

Example 4

Evaluate the following improper integral of the Second Kind, for which the integrand becomes unbounded at a point in the interval of integration, namely, an endpoint.

• q7 := Int(1/sqrt(4 - x), x = 0..4);

$$q7 := \int_0^4 \frac{1}{\sqrt{4-x}}\,dx$$

Since the theory of Riemann integration does not allow infinite integrands, let us try to integrate in such a way as to avoid having the integrand become unbounded. We do this by integrating to something short of x = 4.

• q8 := int(1/sqrt(4 - x), x = 0..4 - z);

$$q8 := -2\sqrt{z} + 4$$

Limits are certainly within our purview, so we will allow z to approach zero through positive values and assign whatever results as the meaning of the improper integral in q6.

• limit(q8, z = 0, right);

Example 5

Evaluate the following improper integral of the Second Kind. Note that now the integrand will become unbounded at an interior point of the interval of integration, namely, x = 0. We convert this case into the one of Example 4 by splitting the integral into two, the first over the interval [-1, 0] and the second, over [0, 2].

• q9 := Int(1/x^3, x = -1..2);

$$q9 := \int_{-1}^{2} \frac{1}{x^3} \, dx$$

Let us look at the integral over the interval [0, 2] by integrating just short of the singularity at x = 0, and taking a limit of that result as the lower limit approaches the singularity.

• q10 := int(1/x^3, x = 0 + z .. 2);

$$q10 := -\frac{1}{8} + \frac{1}{2} \frac{1}{z^2}$$

• limit(q10, z = 0, right);

$$\infty$$

This outcome tells us we can stop right here. The improper integral in q8 does not exist. There is an infinite amount of "area" under part of the curve so the improper integral does not exist.

Example 6

Obtain the Cauchy Principal Value (CPV) for the improper integral in Example 5.

The CPV is an attempt to balance the "positive area" against the "negative area" on either side of a singularity like that at x = 0 in q8. This is done by integrating to within the same distance of the singularity on either side, and then letting the stopping points approach the singularity at the same rate. Analytically, it means computing the following integrals and limit.

• q11 := Int(1/x^3, x = -1..0 - z) + Int(1/x^3, x = 0 + z .. 2);

$$q11 := \int_{-1}^{-z} \frac{1}{x^3}\,dx + \int_{z}^{2} \frac{1}{x^3}\,dx$$

- q12 := value(q11);

$$q12 := \frac{3}{8}$$

And the symmetry of the integrand across the singularity causes the parameter z to cancel out before the need for a limit arises. Thus, the CPV for this improper integral is 3/8.

Finally, we show how Maple can obtain the CPV of an improper integral.

- int(1/x^3, x = -1..2, CauchyPrincipalValue);

$$\frac{3}{8}$$

Example 7

Obtain the CPV of the following improper integral of the Second Kind.

- f := 1/(x - 1)/(x - 3)/(x - 5);

$$f := \frac{1}{(x-1)(x-3)(x-5)}$$

- q13 := Int(f, x = 4..6);

$$q13 := \int_{4}^{6} \frac{1}{(x-1)(x-3)(x-5)}\,dx$$

This time the integrand is not symmetric about the interior singularity at x = 5.

- plot(f, x = 4..6, -5..5);

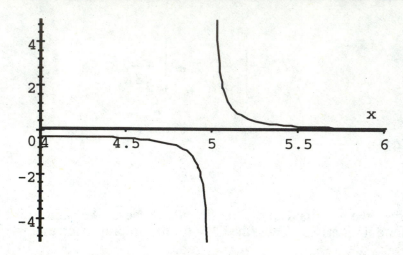

It is unfortunate that Maple integrates $1/x$ to $\ln(x)$ instead of $\ln|x|$. This will generate some extra work for us, but it will be worthwhile, even though we must insert the absolute values ourselves.

- q14 := int(f, x);

$$q14 := \frac{1}{8}\ln(x-1) - \frac{1}{4}\ln(x-3) + \frac{1}{8}\ln(x-5)$$

- q15 := subs(x-1 = abs(x-1), x-3 = abs(x-3), x-5 = abs(x-5), q14);

$$q15 := \frac{1}{8}\ln(|x-1|) - \frac{1}{4}\ln(|x-3|) + \frac{1}{8}\ln(|x-5|)$$

Now the integration

$$\int_{4}^{5-z} \frac{1}{(x-1)(x-3)(x-5)}dx + \int_{5+z}^{6} \frac{1}{(x-1)(x-3)(x-5)}dx$$

can be obtained by the appropriate substitutions into the antiderivative in q15.

- q16 := subs(x = 5 - z, q15) - subs(x = 4, q15)
        + subs(x = 6, q15) - subs(x = 5 + z, q15);

$$q16 := \frac{1}{8}\ln(|4-z|) - \frac{1}{4}\ln(|2-z|) + \frac{1}{8}\ln(|-z|) - \frac{3}{8}\ln(|3|) + \frac{3}{8}\ln(|1|) - \frac{1}{8}\ln(|-1|)$$

$$+ \frac{1}{8}\ln(|5|) - \frac{1}{8}\ln(|4+z|) + \frac{1}{4}\ln(|2+z|) - \frac{1}{8}\ln(|z|)$$

Before we take the limit as z -> 0, it would be nice to tell Maple that z is a small positive quantity and see if Maple could thereby dispense with some of the absolute values.

- assume(z>0);
  additionally(z<1);
  q17 := simplify(q16);

$$q17 := \frac{1}{8}\ln(4 - z\sim) - \frac{1}{4}\ln(2 - z\sim) - \frac{3}{8}\ln(3) + \frac{1}{8}\ln(5) - \frac{1}{8}\ln(4 + z\sim) + \frac{1}{4}\ln(2 + z\sim)$$

- limit(q17, z = 0, right);

$$\frac{1}{8}\ln(5) - \frac{3}{8}\ln(3)$$

We have computed the CPV ourselves, and we end by asking Maple to do the same.

- int(f, x = 4..6, CauchyPrincipalValue);

$$\frac{1}{8}\ln(5) - \frac{3}{8}\ln(3)$$

Exercises

1. Define $f(x) = x^{-3/2}$.

    (a) Plot $f(x)$ for $1 \le x \le 4$.

    (b) Find the area of the region in the first quadrant which is under $f(x)$ and between $x = 1$ and $x = 4$.

    (c) Find the area of the region in the first quadrant which is under $f(x)$ and to the right of $x = 1$. (Find the area of the region between $x = 1$ and $x = b$, and then find the limit as $b \to +\infty$.)

    (d) Find the surface area of the solid formed by revolving $f(x)$, to the right of $x = 1$, about the x-axis.

2. Define $f(x) = \dfrac{1}{\sqrt{x-1}}$.

    (a) Plot $f(x)$ for $1 < x \le 10$.

    (b) Find the area of the region in the first quadrant which is under $f(x)$ and between $x = 1$ and $x = 10$. (Evaluate a proper integral and then find a limit of the result.)

    (c) Here you have a region of infinite extent. Is the size of the area finite or infinite? Does anything here contradict your intuition?

3. Define $f(x) = \exp(x)$.

    (a) Plot $f(x)$ for $-3 \le x \le 1$.

    (b) Find the area of the region in the second quadrant which is under $f(x)$. (You should evaluate a proper integral and then find a limit of your result.)

    (c) Find the volume of the solid formed by revolving the region of part (b) about the x-axis.

4. Define $f(x) = \sqrt{\dfrac{1+x}{1-x}}$.

    (a) Plot $f(x)$ for $0 \le x \le 1$.

    (b) Find the area of the region in the first quadrant under $f(x)$ from $x = 0$ to $x = 1$.

5. Define $f(x) = \frac{\ln x}{x}$ and $g(x) = \frac{\ln x}{x^2}$.

   (a) Plot $f(x)$ and $g(x)$ together for $1 \leq x$.

   (b) Find the area of the region under $f(x)$ in the first quadrant for $1 \leq x$.

   (c) Find the area of the region between $f(x)$ and $g(x)$ in the first quadrant.

6. (a) Print the value of $f(x) = \frac{x^2}{\sin x}$ at $x = 0.5, 0.05, 0.005, 0.0005,$ and 0.00005.

   (b) What do the values in part (a) suggest about $\lim_{x \to 0} f(x)$?

   (c) Answer this limit question using L'Hôpital's Rule.

7. Repeat Exercise 6 for $f(x) = \frac{\sec x}{\tan x}$, using $x = \frac{\pi}{2} - \frac{1}{2}, \frac{\pi}{2} - \frac{1}{4}, \frac{\pi}{2} - \frac{1}{8}, \frac{\pi}{2} - \frac{1}{16}$ and $\frac{\pi}{2} - \frac{1}{32}$ in part (a).

Projects

1. (a) Print the value of $\frac{\sin x}{x}$ at $x_1, ..., x_{10}$, where $x_1 = 0.5$ and $x_{k+1} = \frac{x_k}{2}$ for $k = 1 .. 9$.

   (b) What do the numbers in part (a) suggest about $\lim_{x \to 0} \frac{\sin x}{x}$?

   (c) What does L'Hôpital's Rule say about $\lim_{x \to 0} \frac{\sin x}{x}$?

2. Repeat Exercise 1 for $\lim_{x \to \frac{\pi}{2}} \frac{1 + \tan x}{x}$. (In part (a), you will need to choose $x_1, ..., x_{10}$ "approaching" $\frac{\pi}{2}$.)

3. Repeat Exercise 1 for $\lim_{x \to +\infty} (x + 1)^{\frac{1}{x}}$.

4. Repeat Exercise 1 for $\lim_{x \to +\infty} (1 + \frac{1}{x})^x$. (In part (a), you will need $x_1, ..., x_{10}$ which become large.)

5.   Repeat Exercise 1 for $\lim\limits_{x \to +\infty} (1 + \frac{4}{x^2})^x$. (In part (**a**), you will need $x_1, \dots , x_{10}$ which become large.)

6.   Define $f(x) = \frac{1}{x}$.

   (**a**) Find the area of the region in the first quadrant which is under $f(x)$ from $x = 1$ to $x = b$, for $b = 100$; for $b = 300$; for $b = 900$; for $b = 2700$.

   (**b**) Find the area of the region in the first quadrant which is under $f(x)$ and to the right of $x = 1$.

   (**c**) Find the volume of the solid formed by revolving the region of part (**b**) about the x-axis. (First work with the solid between $x = 1$ and $x = b$.)

   (**d**) Find the surface area of the solid formed by revolving the region of part (**b**) about the x-axis.

   (**e**) This problem works with a region and a solid of infinite extent. What did you determine about the area, volume, and surface area of the objects determined by $f(x)$?

Chapter 11

Sequences and Infinite Series

New Maple Commands for Chapter 11

collect(expression, x);	group like powers of x in expression
convert(taylor(f,x=a,n), polynom);	convert result of **taylor** command to polynomial
taylor(f,x=a,n);	Taylor polynomial of order n - 1, at x = a, for f

---

Introduction

Infinite series are one of the most useful and important objects to emerge from the calculus. The notions of convergence they helped clarify set the tone for modern analysis. Their unique ability to represent and approximate functions makes them the cornerstone of applied mathematics. In particular, the Taylor series is probably the most used notion in the applications. Traditionally, the study of series begins with a study of the convergence of series of numbers. But the graphical capability of a software tool like Maple makes a study of convergence of Taylor series more intuitive, and hence provides a better starting point for thinking about convergence of series. But whatever tack your text takes, the importance of series should not be diminished.

If the outcome of a study of infinite series is comprehension and recollection of the observation that when a series converges, its nth term goes to zero, but not conversely, then that study certainly achieved a measure success.

Examples

Example 1

Show that the sequence $a_n = n^2/2^n$ is monotone decreasing.

One approach to this problem would be treating the expression for a_n as a continuous function and graphing it. We choose, instead, to examine a graph of the ratios a_{n+1}/a_n. If, for each n, a_{n+1} is smaller than a_n then the sequence is monotone decreasing. Finally, if we can solve the inequality $a_{n+1} < a_n$ we might discover analytically for what values of n the sequence is strictly decreasing.

- a := n -> n^2/2^n;

$$a := n \to \frac{n^2}{2^n}$$

• q := seq([n, a(n)], n = 1..15);

$$q := \left[1, \frac{1}{2}\right], [2, 1], \left[3, \frac{9}{8}\right], [4, 1], \left[5, \frac{25}{32}\right], \left[6, \frac{9}{16}\right], \left[7, \frac{49}{128}\right], \left[8, \frac{1}{4}\right], \left[9, \frac{81}{512}\right], \left[10, \frac{25}{256}\right],$$
$$\left[11, \frac{121}{2048}\right], \left[12, \frac{9}{256}\right], \left[13, \frac{169}{8192}\right], \left[14, \frac{49}{4096}\right], \left[15, \frac{225}{32768}\right]$$

• n := 'n':

• plot([q], n = 0..16, 0..3/2, style = point, symbol = circle);

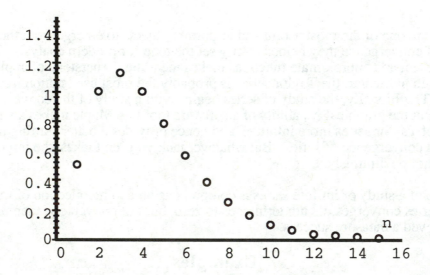

This graph does not contradict an eventual monotone decreasing behavior. What about a graph of ratios of successive terms?

• p := seq([n, a(n+1)/a(n)], n = 1..15);

$$p := [1, 2], \left[2, \frac{9}{8}\right], \left[3, \frac{8}{9}\right], \left[4, \frac{25}{32}\right], \left[5, \frac{18}{25}\right], \left[6, \frac{49}{72}\right], \left[7, \frac{32}{49}\right], \left[8, \frac{81}{128}\right], \left[9, \frac{50}{81}\right], \left[10, \frac{121}{200}\right],$$
$$\left[11, \frac{72}{121}\right], \left[12, \frac{169}{288}\right], \left[13, \frac{98}{169}\right], \left[14, \frac{225}{392}\right], \left[15, \frac{128}{225}\right]$$

• n := 'n':

• plot([p], n = 0..17, 0..1.2, style = point, symbol = circle);

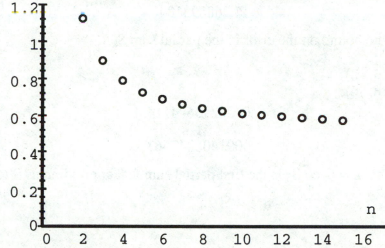

The graph of the ratios seems to indicate that the "next term" remains smaller than the "previous term." We are thereby encouraged to seek analytic evidence of the monotonicity.

• Q := simplify(a(n+1) - a(n));

$$Q := 2^{(-n-1)} n^2 + 2^{(-n)} n + 2^{(-n-1)} - n^2 2^{(-n)}$$

• solve(Q < 0, n);

$$\{ n < 1 - \sqrt{2} \}, \{ \sqrt{2} + 1 < n \}$$

So, for $n \geq 3$ the sequence defined by a_n is monotone decreasing.

Example 2

The alternating series whose kth term a_k is $2/(2 k^4 + 29 k^3 + 124 k^2 + 139 k + 42)$ converges by the alternating series test. Find the smallest value of k for which the kth partial sum S_k, approximates S, the sum of the series, to within .0002.

Since $|S - S_{k-1}| < a_k$, we need to find the value of k for which $|a_k| < .0002$.

• ak := 2/(2*k^4 + 29*k^3 + 124*k^2 + 139*k + 42);

$$ak := 2 \frac{1}{2 k^4 + 29 k^3 + 124 k^2 + 139 k + 42}$$

First, we ask for the value of k for which $a_k < .0002$. Then, the partial sum we want is S_{k-1}.

- fsolve(ak = .0002, k, 1..10);

$$5.268893134$$

The value of a_k is the bound on the error in the partial sum S_{k-1}.

- evalf(subs(k = 5, f));
  evalf(subs(k = 6, f));

$$.0002295684114$$

$$.0001408847563$$

These results say that k = 6, so S_5 is the first partial sum that approximates S to within .0002.

Example 3

When the infinite series having kth term $a_k = [\ln(k)/k]^2$ is approximated by the nth partial sum, the error is less than

$$\int_n^\infty \frac{\ln(x)^2}{x^2}\,dx$$

as is discussed in Exercise 10. Find, in terms of n, a bound for the error. What happens to this bound when n becomes large?

- q := int((ln(x)/x)^2, x = n..infinity);

$$q := \frac{\ln(n)^2 + 2\ln(n) + 2}{n}$$

To show that the bound q goes to zero as n goes to infinity, we take the following limit.

- limit(q, n = infinity);

$$0$$

Example 4

Determine the coefficients of $p(x) = Ax^2 + Bx + C$ so that $p(x)$ and its first two derivatives agree with $\cos(x)$ and *its* first two derivatives at $x = \pi/3$. Plot $p(x)$ and $\cos(x)$ on the same coordinate axes, with $0 \le x \le \pi/2$.

- p := A*x^2 + B*x + C;

$$p := A\,x^2 + B\,x + C$$

- e1 := subs(x = Pi/3, p = cos(x));
  e2 := subs(x = Pi/3, diff(p,x) = -sin(x));
  e3 := subs(x = Pi/3, diff(p,x,x) = -cos(x));

$$e1 := \frac{1}{9}A\,\pi^2 + \frac{1}{3}B\,\pi + C = \cos\left(\frac{1}{3}\pi\right)$$

$$e2 := \frac{2}{3}A\,\pi + B = -\sin\left(\frac{1}{3}\pi\right)$$

$$e3 := 2\,A = -\cos\left(\frac{1}{3}\pi\right)$$

- q := solve({e1, e2, e3}, {A, B, C});

$$q := \left\{ A = \frac{-1}{4}, B = \frac{1}{6}\pi - \frac{1}{2}\sqrt{3}, C = -\frac{1}{36}\pi^2 + \frac{1}{6}\pi\sqrt{3} + \frac{1}{2} \right\}$$

- p1 := subs(q, p);

$$p1 := -\frac{1}{4}x^2 + \left(\frac{1}{6}\pi - \frac{1}{2}\sqrt{3}\right)x - \frac{1}{36}\pi^2 + \frac{1}{6}\pi\sqrt{3} + \frac{1}{2}$$

- plot({p1, cos(x)}, x = 0..Pi/2);

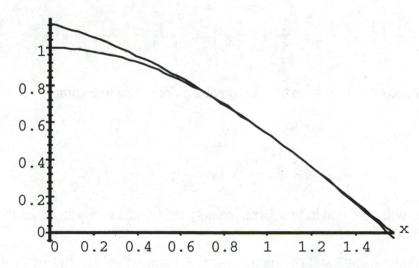

Example 5

At $x = \pi/3$, find p2 and p4, the second- and fourth-degree Taylor polynomials for f(x) = cos(x). Compare p2 to the polynomial found in Example 4. Plot p2, p4, and cos(x) on the same set of axes, with $0 \le x \le \pi$.

- q1 := taylor(cos(x), x = Pi/3, 3);
  q2 := taylor(cos(x), x = Pi/3, 5);

$$q1 := \frac{1}{2} - \frac{1}{2}\sqrt{3}\left(x - \frac{1}{3}\pi\right) - \frac{1}{4}\left(x - \frac{1}{3}\pi\right)^2 + O\left(\left(x - \frac{1}{3}\pi\right)^3\right)$$

$$q2 := \frac{1}{2} - \frac{1}{2}\sqrt{3}\left(x - \frac{1}{3}\pi\right) - \frac{1}{4}\left(x - \frac{1}{3}\pi\right)^2 + \frac{1}{12}\sqrt{3}\left(x - \frac{1}{3}\pi\right)^3 + \frac{1}{48}\left(x - \frac{1}{3}\pi\right)^4 +$$
$$O\left(\left(x - \frac{1}{3}\pi\right)^5\right)$$

The "Big Oh" terms indicate the order of the truncation error. Until these error terms are removed, you cannot manipulate the result of the **taylor** command.

- p2 := convert(q1, polynom);
  p4 := convert(q2, polynom);

$$p2 := \frac{1}{2} - \frac{1}{2}\sqrt{3}\left(x - \frac{1}{3}\pi\right) - \frac{1}{4}\left(x - \frac{1}{3}\pi\right)^2$$

$$p4 := \frac{1}{2} - \frac{1}{2}\sqrt{3}\left(x - \frac{1}{3}\pi\right) - \frac{1}{4}\left(x - \frac{1}{3}\pi\right)^2 + \frac{1}{12}\sqrt{3}\left(x - \frac{1}{3}\pi\right)^3 + \frac{1}{48}\left(x - \frac{1}{3}\pi\right)^4$$

To compare p2 computed here with the polynomial p1 obtained in Example 4, we need to **collect** like powers in p2.

- collect(p2, x);

$$-\frac{1}{4}x^2 + \left(\frac{1}{6}\pi - \frac{1}{2}\sqrt{3}\right)x - \frac{1}{36}\pi^2 + \frac{1}{6}\pi\sqrt{3} + \frac{1}{2}$$

The astute reader will note that in this form, the polynomial p2 exactly matches the polynomial p1 found in Example 4.

A plot of the polynomials p2 and p4 against cos(x) illustrates the behavior of Taylor polynomials.

• plot({p2, p4, cos(x)}, x = 0..Pi);

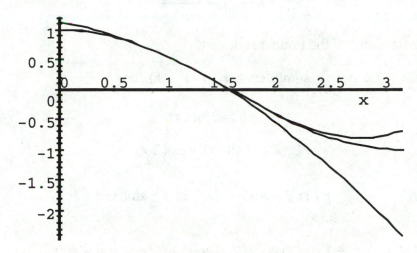

Example 6

Examine the series

$$\sum_{n=1}^{\infty} b_n \sin(n\,x)$$

which represents the function f(x) = x, $0 \le x \le \pi$, if the coefficients b_n are computed as

$$b_n = 2 \frac{\displaystyle\int_0^{\pi} f(x)\,dx}{\pi}$$

Such a series is called a Fourier series. It is fitting to note the existence of such series whose coefficients are determined by integration rather than by differentiation as in a Taylor series.

• b := (2/Pi)*int(x*sin(n*x), x = 0..Pi);

$$b := -2\,\frac{-\sin(n\,\pi) + n\cos(n\,\pi)\,\pi}{\pi\,n^2}$$

Since $\sin(n\pi) = 0$ for n an integer, we will remove it from the expression for b by brute force.

• b := subs(sin(n*Pi) = 0, b);

$$b := -2 \frac{\cos(n\,\pi)}{n}$$

The first five partial sums of the Fourier series are

- for k from 1 to 5 do p.k := sum(b*sin(n*x), n = 1..k); od;

$$p1 := 2\sin(x)$$

$$p2 := 2\sin(x) - \sin(2\,x)$$

$$p3 := 2\sin(x) - \sin(2\,x) + \frac{2}{3}\sin(3\,x)$$

$$p4 := 2\sin(x) - \sin(2\,x) + \frac{2}{3}\sin(3\,x) - \frac{1}{2}\sin(4\,x)$$

$$p5 := 2\sin(x) - \sin(2\,x) + \frac{2}{3}\sin(3\,x) - \frac{1}{2}\sin(4\,x) + \frac{2}{5}\sin(5\,x)$$

- plot({p5, x}, x = 0..Pi);

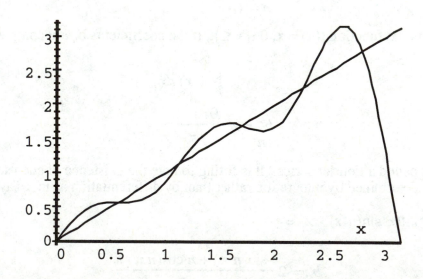

The utility and importance of Fourier series are second only to those of Taylor series. No study of classical applied mathematics is complete without a good look at the Fourier series and its generalizations to other orthogonal functions. See Exercises 20 - 31, as well as Problems 15 - 24, for further insights into the Fourier series.

Exercises

1. Define each sequence using the Maple command **seq**; then plot several terms of the sequence as a collection of points. Predict whether the sequence is convergent or divergent.

    (a) $\{1 - \frac{1}{n}\}_{n=1}^{\infty}$ **(c)** $\{\frac{2n+1}{2^n}\}_{n=1}^{\infty}$

    (b) $\{\frac{2n+1}{n+1}\}_{n=1}^{\infty}$ **(d)** $\{\frac{n!}{5^n}\}_{n=1}^{\infty}$.

2. Define each sequence using the Maple command **seq**; then plot several terms of the sequence as a collection of points. Can you predict from the plot whether the sequence is monotone increasing or decreasing from some point on? To confirm your prediction, test $\dfrac{a_{n+1}}{a_n}$.

    (a) $\{\frac{2n-1}{n+1}\}_{n=1}^{\infty}$ **(d)** $\{\frac{n!}{4^n}\}_{n=1}^{\infty}$

    (b) $\{\frac{2^n-1}{2^n+1}\}_{n=1}^{\infty}$ **(e)** $\{\frac{2^n}{(2n)!}\}_{n=1}^{\infty}$.

    (c) $\{\frac{3n}{3^n}\}_{n=1}^{\infty}$

3. Define three sequences by $a_n = \frac{1}{10}n^2$, $b_n = \frac{1}{10}2^n$, and $c_n = \frac{1}{10}n!$. Which of the three do you think will eventually outgrow the other two? Which will grow at the slowest rate as $n \to \infty$? In the same window, plot the three functions that match the sequences for integer x values. (To distinguish between the functions, you could assign each plot a different thickness or color.)

4. In the same window, plot the sequence and the sequence of partial sums for n = 1..20. For the sequence use 'symbol = BOX', and for the sequence of partial sums use 'symbol = DIAMOND'. From these plots decide whether the sequence of partial sums (and the resulting infinite series) converges or diverges.

    (a) $\{\frac{1}{2^n}\}_{n=1}^{\infty}$ **(c)** $\{\frac{n}{3}\}_{n=1}^{\infty}$

    (b) $\{\frac{1}{n}\}_{n=1}^{\infty}$ **(d)** $\{\frac{2}{n^2}\}_{n=1}^{\infty}$.

5. Find the smallest value of n for which $\displaystyle\sum_{k=1}^{n} \frac{1}{k}$ is greater than   (a) 5   (b) 10.

    (c) Compare your answer in part (b) with the result of grouping approximately that number of terms in the form $1 + \frac{1}{2} + (\frac{1}{3} + \frac{1}{4}) + (\frac{1}{5} + \cdots + \frac{1}{8}) + \cdots$.

6. For the given nth partial sum $s_n = \displaystyle\sum_{k=1}^{n} a_k$, of $\displaystyle\sum_{k=1}^{\infty} a_k$, find a_k and $\displaystyle\sum_{k=1}^{\infty} a_k$.

    (a) $s_n = \dfrac{2n}{2n-1}$ (b) $s_n = \dfrac{n-1}{n+1}$.

7. A bicycle rolling down a long section of level road loses 8% of its speed every 500 meters. If the initial speed was 30 mph, what is its speed after 2500 meters? After 5000 meters? When does its speed reach 1 mph? When does its speed become 0?

8. The harmonic series $\displaystyle\sum_{k=1}^{\infty} \frac{1}{k}$ diverges to ∞, but very slowly. Suppose that 10 billion years ago, some of our ancestors started with the first term in the harmonic series, and each day since, 10,000 terms have been added. What would be the approximate present value of $\displaystyle\sum_{k=1}^{n} \frac{1}{k}$? Compare your answer with the approximation given by the following inequality: $\ln(n+1) \le \displaystyle\sum_{k=1}^{n} \frac{1}{k} \le 1 + \ln n$.

9. The infinite series $\displaystyle\sum_{k=1}^{\infty} \frac{1}{k^2}$ is known to converge. Therefore, the nth partial sum gives an approximation to the sum of the series, and the error in using such an approximation is $\displaystyle\sum_{k=n+1}^{\infty} \frac{1}{k^2}$. Find a bound for this error by treating $\displaystyle\sum_{k=n+1}^{\infty} \frac{1}{k^2}$ as a sum of areas of rectangles, each having a base of length 1, and comparing with the area under the curve $f(x) = \dfrac{1}{x^2}$. Can you use Maple directly to compute the number of terms of $\displaystyle\sum_{k=1}^{\infty} \frac{1}{k^2}$

needed for a two decimal-place value?

10. If $\sum\limits_{k=1}^{\infty} u_k$ is an infinite series that is convergent by the Integral Test, and the sum is

approximated by the nth partial sum S_n, then the error is less than $\int_n^\infty f(x)\,dx$. ($f(x)$ is

related to u_k as in the Integral Test.)

For each of the following infinite series that converge by the Integral Test, find a bound for the error when the sum is approximated by the nth partial sum.

(a) $\sum\limits_{k=1}^{\infty} \dfrac{1}{k^2}$, $n = 50$ $\qquad$ (e) $\sum\limits_{k=1}^{\infty} \dfrac{k}{k+1}$, $n = 100$

(b) $\sum\limits_{k=1}^{\infty} \dfrac{1}{k^3}$, $n = 50$ $\qquad$ (f) $\sum\limits_{k=1}^{\infty} \dfrac{\ln k}{k^2}$, $n = 50$

(c) $\sum\limits_{k=1}^{\infty} \dfrac{1}{e^k}$, $n = 10$ $\qquad$ (g) $\sum\limits_{k=1}^{\infty} \dfrac{k}{e^k}$, $n = 50$.

(d) $\sum\limits_{k=1}^{\infty} \dfrac{\ln k}{k}$, $n = 100$

11. The series $\sum\limits_{k=1}^{\infty} (-1)^{k+1} \dfrac{k}{4+k^2}$ converges by the Alternating Series Test.

(a) Find (as a decimal value) $\sum\limits_{k=1}^{60} (-1)^{k+1} \dfrac{k}{4+k^2}$

(b) If your answer in part (a) is used as an estimate for $\sum\limits_{k=1}^{\infty} (-1)^{k+1} \dfrac{k}{4+k^2}$, what is the

maximum $|\text{Error}|$?

(**c**) Find the smallest positive integer value n such that $\sum_{k=1}^{n}(-1)^{k+1}\dfrac{k}{4+k^2}$ approximates

the infinite series with $|\text{Error}| \leq 0.005$.

12. What value of n is required in order that $\sum_{k=1}^{n}\dfrac{(-1)^{k+1}}{k^2+3k+2}$ approximates the sum of the

infinite alternating series $\sum_{k=1}^{\infty}\dfrac{(-1)^{k+1}}{k^2+3k+2}$ to four decimal-place accuracy?

13. The series $\sum_{k=1}^{\infty}\dfrac{(-1)^k}{k^3+k^2+10k-6}$ converges by the Alternating Series Test. Find the smallest

value of n for which the nth partial sum approximates the sum of the series with $|\text{Error}|$ < 0.000001. Is this approximation too large or too small?

14. Define $f(x) = \tan(x)$.

(**a**) At the point x = 0, find p3(x), the third-degree Taylor polynomial for f(x). (Note that this is a Maclaurin polynomial.)

(**b**) Estimate the value of f(0.5) by finding p3(0.5). Likewise, estimate f(1.0).

(**c**) Print the error in p3(0.5). (That is, print p3(0.5) - f(0.5).) Likewise, print the error in p3(1.0).

(**d**) In the same window, plot p3(x) and f(x) for $0 \leq x \leq 1.4$.

(**e**) On what interval for x does p3(x) appear to be a good approximation to f(x)?

(**f**) If you are asked to estimate f(0.4), and can use only pencil and paper, note that p3(x) gives you a reasonable solution. This type of approach also gives an effective means to find the value of functions such as e^x, ln(x), arcsin(x), and so on.

(**g**) Find p5(x), the degree five Taylor polynomial for f(x). In the same window, plotf(x), p3(x), and p5(x) for $0 \leq x \leq 1.4$.

(**h**) Neither p3(x) nor p5(x) gives an acceptable value of f(1.3). How can you modify your Taylor polynomial to improve your estimate of f(1.3)? What is your method if you are not allowed to increase the degree of your Taylor polynomial?

15. Define $f(x) = \sin(x)$. Repeat each part of Exercise 14 for this function.

16. Find the Taylor polynomial of the specified degree for f(x) at x = a. Then plot your Taylor polynomial, along with f(x) on the given interval. Does higher degree appear to give a better fit to f(x)?

   (a) $f(x) = \cos(x)$,   a = 0,   p2 and p4,   on $[0, \pi]$

   (b) $f(x) = \tan(x)$, $a = \frac{\pi}{4}$,   p1 and p3,   on [-0.5, 2]

   (c) $f(x) = \exp(x)$,   a = 0.5,   p2 and p3,   on [-1, 2]

   (d) $f(x) = \sqrt{x^2 + 1}$,   a = 1,   p2 and p3,   on [0, 2].

17.   Define $t_p(n)$ to be the nth degree Taylor polynomial of sin(x) about $x = \frac{\pi}{4}$.

   (a) Make a movie in which each frame shows the plot of sin(x) and $t_p(k)$ on [0..3], for k = 2..6.

   (b) Repeat part (a) on [0..4] for k = 2..9.

18. For each of the following definite integrals, obtain a numeric value by using Maple's built-in numeric integrator. Then replace each integrand with a Taylor polynomial of degree ≤ 4. Evaluate these new definite integrals exactly, and then approximately via Simpson's Rule with n = 4. Compare the values found through these various techniques.

   (a) $\int_0^{0.3} \sqrt{1 + x^3}\ dx$ (d) $\int_0^{0.2} \tan(x^2)\ dx$

   (b) $\int_{1.8}^{2.1} \sqrt{1 + x^3}\ dx$ (e) $\int_{1.5}^{2} \sin(\sqrt{x})\ dx.$

   (c) $\int_0^{0.4} \sqrt{1 + x^4}\ dx$

19. If f(x) is approximated by the Taylor polynomial $p_n(x)$ at x = a, what is the maximum error on the specified interval?

   (a) $f(x) = \sin(x) \cos(2x)$,   p2(x),   a = 0,   on [0, 0.5]

   (b) $f(x) = \sin(x) \ln(x)$,   p3(x),   a = 1,   on [1, 1.4]

   (c) $f(x) = \exp(-\frac{x}{2}) \cos(2x)$,   p3(x),   a = 0,   on [0, 1]

(d) $f(x) = x^2 exp(-x)$,   p6(x), $a = 0$,   on $[0, 0.5]$

(e) $f(x) = \sinh(x)$,   p5(x), $a = 0$,   on $[0, 1]$.

20. Find the smallest value of n for which the Taylor polynomial, $p_n(x)$, is accurate to m decimal places throughout the given interval.

   (a) $f(x) = \sin(x) \cos(2x)$, $a = 0$,   on $[0, 0.5]$, $m = 2$

   (b) $f(x) = \sin(x) \ln(x)$, $a = 1$,   on $[1, 1.4]$, $m = 4$

   (c) $f(x) = exp(-\frac{x}{2}) \cos(2x)$, $a = 0$,   on $[0, 1]$, $m = 2$

   (d) $f(x) = x^2 exp(-x)$, $a = 0$,   on $[0, 0.5]$, $m = 5$

   (e) $f(x) = \sinh(x)$, $a = 0$,   on $[0, 1]$, $m = 4$.

21. Obtain the Fourier sine series for $f(x) = x^2$, $0 \le x \le \pi$. Plot p20 on the same three intervals as in Exercise 20. Describe the function to which each Fourier series converges.

22. Repeat Exercise 21 for $f(x) = \begin{cases} 0, & 0 \le x \le \frac{\pi}{2} \\ x - \frac{\pi}{2}, & \frac{\pi}{2} \le x \le \pi \end{cases}$.

A Fourier cosine series for $f(x)$, $0 \le x \le \pi$, is given by the sum $\dfrac{a_0}{2} + \displaystyle\sum_{n=1}^{\infty} a_n \cos(nx)$, with

$a_n = \dfrac{2}{\pi} \displaystyle\int_0^{\pi} f(x) \cos(nx)\, dx$. For each function below, find the Fourier cosine series and plot p20 on the same three intervals given in Exercise 21. Again, in each case, describe the function to which the Fourier series converges.

23. The function $f(x) = x$, $0 \le x \le \pi$.

24. The function $f(x) = x^2$, $0 \le x \le \pi$.

25. The function of Exercise 22.

A Fourier series for $f(x)$, $-\pi \le x \le \pi$, containing both sine and cosine terms is given by the sum

$\dfrac{a_0}{2} + \displaystyle\sum_{n=1}^{\infty}(a_n \cos(nx) + b_n \sin(nx))$, with $a_n = \dfrac{1}{\pi}\displaystyle\int_{-\pi}^{\pi} f(x) \cos(nx)\, dx$, $b_n = \dfrac{1}{\pi}\displaystyle\int_{-\pi}^{\pi} f(x) \sin(nx)\, dx$.

For each function below, find the Fourier series and plot p20 on the three intervals mentioned in Exercise 21.

26. The function $f(x) = x$, $-\pi \le x \le \pi$.

27. The function $f(x) = x^2$, $-\pi \le x \le \pi$.

28. The function $f(x) = \begin{cases} 1, & -\pi \le x < 0 \\ -1, & 0 \le x \le \pi \end{cases}$

29. The function $f(x) = \begin{cases} 1, & -\pi \le x < 0 \\ x, & 0 \le x \le \pi \end{cases}$

30. The function $f(x) = \begin{cases} \pi^2 - x^2, & -\pi \le x \le 0 \\ \pi(\pi - x), & 0 \le x \le \pi \end{cases}$

31. The function $f(x) = \begin{cases} x(\pi - x), & -\pi \le x \le 0 \\ x, & 0 \le x \le \pi \end{cases}$

Projects

1. In this problem, if you cannot solve an inequality, solve the corresponding equality and work from that. If you cannot solve an equality, use a plot and improvise.

   (a) Define $a_n = \frac{2n}{n+1}$. Since $\lim\limits_{n \to +\infty} a_n = 2$, for $\varepsilon > 0$, there is a positive integer N such that $\left| \frac{2n}{n+1} - 2 \right| < \varepsilon$ for all $n \ge N$. Find the smallest such N when $\varepsilon = 0.01$; when $\varepsilon = 0.0005$; when $\varepsilon = \varepsilon$.

   (b) Repeat part (a) with $a_n = \sum\limits_{k=1}^{n} \frac{2}{3^k}$. Note that in this problem you are working with an infinite series through its sequence of partial sums.

2. Define $a_k = \frac{1}{k}$ and $b_k = \frac{1}{k^2}$. Define $s1_n = \sum\limits_{k=1}^{n} a_k$ and $s2_n = \sum\limits_{k=1}^{n} b_k$. In the same window, plot the sequences $s1_n$ and $s2_n$ for $n = 1..20$. For the sequence $s1_n$ use 'symbol = BOX',

and for $s2_n$ use 'symbol = DIAMOND'. What familiar infinite series is seen to converge, and what familiar infinite series is seen to diverge by your plot?

3. Let $\{C_k\}$, k = 1, 2, ..., be a sequence of circles, each having center at (0, 0), with circle C_k having radius $= \frac{1}{k}$. Find an approximation to $\sum\limits_{k=1}^{\infty} A_k$, where A_1 is the area of the region in quadrant 1 which is inside of C_1 and outside of C_2; A_2 is the area of the region in quadrant 2 which is inside of C_2 and outside of C_3, and so on. Plot the first ten regions.

4. Each infinite series $\sum\limits_{k=1}^{\infty} u_k$ below is convergent by the Integral Test, so the error in the *n*th partial sum is less than $\int_n^{\infty} f(x)\, dx$, where f(x) is related to u_k as in the Integral Test. Display the regions that represent the error and the integral bound of the error, on the interval [n, b].

(a) $\sum\limits_{k=1}^{\infty} \frac{1}{k^2}$; [10, 20]   (b) $\sum\limits_{k=1}^{\infty} \frac{\ln k}{k^2}$; [10, 20]   (c) $\sum\limits_{k=1}^{\infty} \frac{1}{e^k}$; [4, 9].

5. Decide whether each of the infinite series below is likely to converge or diverge. Then choose an infinite series for which you can compare the one being tested, and plot both (the sequences of partial sums) on an appropriate domain. Finally, confirm your comparison algebraically.

(a) $\sum\limits_{k=1}^{\infty} \frac{k}{k^2-1}$ (b) $\sum\limits_{k=1}^{\infty} \frac{1}{k^2+1}$ (c) $\sum\limits_{k=1}^{\infty} \frac{1}{k^2-k+3}$.

6. Define $f(x) = \cos(x)$, and $p2(x) = ax^2 + bx + c$.

(a) Determine a, b, and c so that $p2(\frac{1}{2})$, $p2'(\frac{1}{2})$, and $p2''(\frac{1}{2})$ agree, respectively, with $f(\frac{1}{2})$, $f'(\frac{1}{2})$, and $f''(\frac{1}{2})$.

(b) Plot f(x) and p2(x) for $-\pi \le x \le \pi$.

(c) Where does the best fit occur? What would you do differently in part (a) if you wanted a good fit for $-1 \le x \le 0$?

7. Define $f(x) = \cos(x)$, and $p4(x) = a\,x^4 + b\,x^3 + c\,x^2 + d\,x + e$. Repeat the different parts of Exercise 6, with modifications as needed to reflect the higher degree of p4(x). Then plot f(x), p2(x), and p4(x) for $-\pi \le x \le \pi$. Which of p2(x), p4(x) is a better fit to f(x)?

8. Use the command **taylor** to obtain the degree two and degree four Taylor polynomials of $f(x) = \cos(x)$ at $x = \frac{1}{2}$. Compare with p2(x) and p4(x) from the previous two problems. Will the process of finding coefficients as in Exercises 6 and 7 always yield the Taylor polynomial?

9. Find the fifth-degree Maclaurin polynomial p5(x) for the function e^x. Then find $\int p5(x)\,dx + C$ and compare your answer with the sixth-degree Maclaurin polynomial p6(x) for an appropriate choice of C. What does this suggest about $\int e^x\,dx$?

10. The function f(x) will have an nth degree Maclaurin polynomial $p_n(x)$ with Lagrange Remainder $R_n(x) = f^{(n+1)}(c)\dfrac{x^{n+1}}{(n+1)!}$, where $f^{(k)} = \dfrac{d^k f}{dx^k}$, and c is (a generally unknown point) in the interval [0, x]. For the function $f(x) = e^x$, $0 \le x \le 1$, estimate the maximum value of $R_3(x)$. This maximum will occur where $f'''(c)$ is itself a maximum.

11. Repeat Exercise 10 for the function $\cos(x)$. Use the fourth-degree Maclaurin polynomial and estimate the maximum error for $0 \le x \le \frac{\pi}{3}$.

12. How many pairs of rabbits can be produced from a single pair in a year, if every month each pair begets a new pair, which from the second month on becomes productive?
    -from "An Introduction to the History of Mathematics", third edition, by Howard Eves.

    Hint: At the end of the first month there is one pair, at the end of the second month there is still one pair, at the end of the third month there are two pairs, and so on; thus obtain a sequence $a_n = \{1, 1, 2, ...\}$.

    (a) List the first seven terms of the sequence.

    (b) How is a_{n+2} related to a_n and a_{n+1} for n = 1; for n = 2; for n = 3? Define a_{n+2} in terms of a_n and a_{n+1}.

    (c) Set rabbiteq := {a(1) = 1, a(2) = 1, a(n + 2) = ??} and then use the command **rsolve** to find a solution to the problem.

    (d) How many pairs of rabbits are present at the end of 12 months? 24 months?

    (e) This is a very famous sequence, named after the twelfth-century scholar Leonardo Fibonacci. You can check your answers to part (d) by using the Maple procedure

fibonacci which is found in the package "combinat."

13. Write Maple instructions for performing the "Divergence Test" on an infinite series. Given the kth term of an infinite series, Maple should respond with "the series diverges" or "the test does not apply."

14. Write Maple instructions for performing the "Ratio Test" on an infinite series.

15. The Fourier sine series $\sum_{n=1}^{\infty} b_n \sin(nx)$, $b_n = \frac{2}{\pi} \int_0^{\pi} f(x) \sin(nx)\, dx$, represents a function $f(x)$ defined on the interval $0 \le x \le \pi$. Show that the functions $\varphi_n(x) = \sqrt{\frac{2}{\pi}} \sin(nx)$, $n = 1, 2, \ldots$ are "orthonormal" in the sense that $\int_0^{\pi} \varphi_n(x)\varphi_m(x)\, dx = \begin{cases} 1, \text{ if } n = m \\ 0, \text{ if } n \ne m \end{cases}$.

16. The Fourier cosine series $\frac{a_0}{2} + \sum_{n=1}^{\infty} a_n \cos(nx)$, $a_n = \frac{2}{\pi} \int_0^{\pi} f(x) \cos(nx)\, dx$, represents a function $f(x)$ defined on the interval $0 \le x \le \pi$. Show that the functions $\varphi_n(x) = \sqrt{\frac{2}{\pi}} \cos(nx)$, $n = 0, 1, 2, \ldots$ are "orthonormal" in the sense that $\int_0^{\pi} \varphi_n(x)\varphi_m(x)\, dx = \begin{cases} 1, \text{ if } n = m \\ 0, \text{ if } n \ne m \end{cases}$.

17. Show that the polynomial functions $\varphi_n(x) = x^n$, $n = 0, 1, 2, \ldots$, which are the "basis" functions of the Taylor series, are not orthogonal, in the sense of Problems 15 and 16, on any finite interval [a,b].

18. The infinite set of Legendre polynomials $P_0(x) = 1$, $P_1(x) = x$, $P_2(x) = (3x^2 - 1)/2$, $P_3(x) = x(5x^2 - 3)/2$, ... are orthogonal with respect to integration on $-1 \le x \le 1$ and satisfy $\int_{-1}^{1} P_k(x)^2 dx = \frac{2}{2k+1}$. They can be accessed in Maple by first loading the orthopoly package via with(orthopoly) and entering P(k,x) where P(k,x) is $P_k(x)$. Obtain the Legendre polynomials $P_0(x), \ldots, P_5(x)$, and plot them all on the same set of axes for $-1 \le x \le 1$.

An arbitrary but "nice" function $f(x)$ can be expanded in a series of Legendre polynomials $f(x)$

$$f(x) = \sum_{n=0}^{\infty} p_n P_n(x), \text{ with } p_n = \frac{2}{2n+1} \int_{-1}^{1} f(x)\, P_n(x)\, dx.$$ Such a series is called a Fourier-Legendre series. For each function $f(x)$ below, obtain s_{10}, the partial sum of the Fourier-Legendre series up through $P_{10}(x)$, and plot s_{10} and $f(x)$ on the same set of axes for $-1 \le x \le 1$.

19. The function $f(x) = \sin(x)$, $-1 \le x \le 1$.

20. The function $f(x) = e^x$, $-1 \le x \le 1$.

21. The function $f(x) = \dfrac{1}{1 + x^2}$, $-1 \le x \le 1$.

22. The function $f(x) = \begin{cases} 1, & -1 \le x < 0 \\ -1, & 0 \le x \le 1 \end{cases}$

23. The function $f(x) = \begin{cases} 1, & -1 \le x < 0 \\ x^2, & 0 \le x \le 1 \end{cases}$

24. The function $f(x) = \begin{cases} x(1 - x), & -1 \le x \le 0 \\ x, & 0 \le x \le 1 \end{cases}$

Chapter 12

Analytic Geometry

New Maple Commands for Chapter 12

coeff(expression,x,2);	coefficient of x^2 in expression
combine(expression,trig);	perform trigonometric compression on expression
with(student); completesquare(expr,[x,y]);	loads student package completes square in x and y in expression expr

---

Introduction

Chapter 12 contains examples designed to make the conic sections familiar enough that they become useful in examples and constructions in calculus. These curves are among the oldest known in mathematics and are still elegant, if not functional. A typical way of studying conics is by starting from the quadratic equation defining them, then analyzing and plotting. Another way is to synthesize the defining equation that satisfies prescribed conditions. Throughout, understand that it is not so much the conics that are important as is the underlying familiarity with analytic geometry. And this familiarity, with an accompanying facility for understanding and using curves in coordinate geometry, is the distillation of examples such as those presented below.

Examples

Example 1

Find the center and foci of the conic described by $9 x^2 + 4 y^2 + 36 x - 24 y + 36 = 0$.

Since it is so easy to do in Maple, begin with a graph. Note that the given equation defines y(x) implicitly so we either have to solve for y(x) explicitly or use Maple's **implicitplot** command. We opt for the latter alternative, remembering that the **implicitplot** command is in the plots package.

• with(plots):

• q := 9*x^2 + 4*y^2 + 36*x - 24*y + 36 = 0;

$$q := 9 x^2 + 4 y^2 + 36 x - 24 y + 36 = 0$$

• implicitplot(q, x = -5..1, y = -1..7, scaling = constrained);

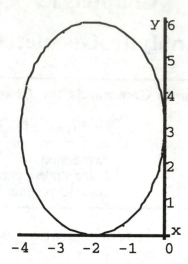

The analysis is eased if we put the quadratic into standard form by completing the square.

- with(student):
- q1 := completesquare(q, [x, y]);

$$q1 := 4\,(y-3)^2 - 36 + 9\,(x+2)^2 = 0$$

Next, put this ellipse into standard form. Add 36 to both sides, then divide through by 36 to achieve a "1" on the right-hand side.

- q2 := q1 + (36 = 36);

$$q2 := 4\,(y-3)^2 + 9\,(x+2)^2 = 36$$

- q3 := q2/36;

$$q3 := \frac{1}{9}(y-3)^2 + \frac{1}{4}(x+2)^2 = 1$$

The center is at (-2, 3), a = 3, b = 2, c = $\sqrt{(a^2 - b^2)}$ = $\sqrt{5}$, and the foci are at (-2, 3 ± c) = (-2, 3 ± $\sqrt{5}$).

Example 2

Find the equation of the conic that passes through the following five points: (0, 1), (2, 3), (1, 0), (-2, 0), (0, 3). Note that it takes five points to determine a conic. Graph the resulting conic and analyze it.

The general form of a conic is the complete quadratic given by

- q4 := a*x^2 + b*x*y + c*y^2 + d*x + e*y + f = 0;

$$q4 := a\,x^2 + b\,x\,y + c\,y^2 + d\,x + e\,y + f = 0$$

There are six constants in q4, but at least one of them must be nonzero. In that case, divide through by the nonzero constant and rename the five ratios so formed. These are then five unknowns, so it takes just five points to determine a conic.

Enter the five given points.

• p1 := [0,1]; p2 := [2, 3]; p3 := [1, 0]; p4 := [-2, 0]; p5 := [0, 3];

$$p1 := [0, 1]$$

$$p2 := [2, 3]$$

$$p3 := [1, 0]$$

$$p4 := [-2, 0]$$

$$p5 := [0, 3]$$

Form five equations in the six unknowns a, b, c, d, e, f by substituting the points p1, ..., p5 into the quadratic q4.

• for k from 1 to 5 do   e.k := subs(x = p.k[1], y = p.k[2], q4);   od;

$$e1 := c + e + f = 0$$

$$e2 := 4\,a + 6\,b + 9\,c + 2\,d + 3\,e + f = 0$$

$$e3 := a + d + f = 0$$

$$e4 := 4\,a - 2\,d + f = 0$$

$$e5 := 9\,c + 3\,e + f = 0$$

Have Maple solve this set of equations, letting Maple decide which of the six unknowns to make arbitrary.

• q5 := solve({e.(1..5)}, {a,b,c,d,e,f});

$$q5 := \left\{ d = -\frac{3}{2}c, a = -\frac{3}{2}c, b = \frac{3}{2}c, e = -4\,c, f = 3\,c, c = c \right\}$$

Maple took c as arbitrary. The quadratic in q4 is now

• q6 := subs(q5, q4);

$$q6 := -\frac{3}{2}c\,x^2 + \frac{3}{2}c\,x\,y + c\,y^2 - \frac{3}{2}c\,x - 4\,c\,y + 3\,c = 0$$

Setting $c = 2$ will remove fractions.

- q7 := subs(c = 2, q6);

$$q7 := -3\,x^2 + 3\,x\,y + 2\,y^2 - 3\,x - 8\,y + 6 = 0$$

Obtain a graph by using the **implicitplot** command.

- implicitplot(q7, x = -4..4, y = -4..6);

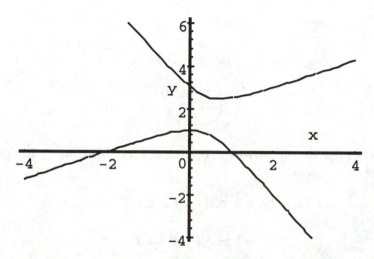

The conic is a rotated hyperbola. The center and foci follow from standard form, obtained by rotating the conic so its axes are parallel to the coordinate axes. We begin by rotating the frame of the hyperbola about the origin in the xy-plane. The angle of rotation is determined by the equation $\cot(2s) = (a - c)/b$, where a, b, and c are coefficients of x^2, xy and y^2, respectively, in the conic to be rotated. Thus, a = -3, b = 3, c = 2 so $\tan(2s) = 3/(-5)$ and $s = -\arctan(3/5)/2$.

- s := -arctan(3/5)/2;

$$s := -\frac{1}{2}\arctan\!\left(\frac{3}{5}\right)$$

The formulas for rotating the plane through angle s are

- q8 := {x = X*cos(s) - Y*sin(s), y = X*sin(s) + Y*cos(s)};

$$q8 := \{\, y = -X\sin(\%1) + Y\cos(\%1),\, x = X\cos(\%1) + Y\sin(\%1)\,\}$$

$$\%1 := \frac{1}{2}\arctan\!\left(\frac{3}{5}\right)$$

Make this rotation in the hyperbola q7.

- q9 := subs(q8, q7);

$$q9 := -3\,(X\cos(\%1) + Y\sin(\%1))^2$$
$$+ 3\,(X\cos(\%1) + Y\sin(\%1))\,(-X\sin(\%1) + Y\cos(\%1))$$
$$+ 2\,(-X\sin(\%1) + Y\cos(\%1))^2 - 3\,X\cos(\%1) - 3\,Y\sin(\%1) + 8\,X\sin(\%1)$$
$$- 8\,Y\cos(\%1) + 6 = 0$$
$$\%1 := \frac{1}{2}\arctan\!\left(\frac{3}{5}\right)$$

Clear parentheses by

- q10 := expand(q9);

$$q10 := -3\,X^2\cos(\%1)^2 - 10\,X\cos(\%1)\,Y\sin(\%1) - 3\,Y^2\sin(\%1)^2$$
$$- 3\,X^2\cos(\%1)\sin(\%1) + 3\,X\cos(\%1)^2\,Y - 3\,Y\sin(\%1)^2\,X$$
$$+ 3\,Y^2\sin(\%1)\cos(\%1) + 2\,X^2\sin(\%1)^2 + 2\,Y^2\cos(\%1)^2 - 3\,X\cos(\%1)$$
$$- 3\,Y\sin(\%1) + 8\,X\sin(\%1) - 8\,Y\cos(\%1) + 6 = 0$$
$$\%1 := \frac{1}{2}\arctan\!\left(\frac{3}{5}\right)$$

Compress some of the trigonometry by

- q11 := combine(q10, trig);

$$q11 := -\frac{1}{2}X^2\sqrt{34} - \frac{1}{2}X^2 - \frac{1}{2}Y^2 + \frac{1}{2}Y^2\sqrt{34} - 3\,X\cos(\%1) - 3\,Y\sin(\%1)$$
$$+ 8\,X\sin(\%1) - 8\,Y\cos(\%1) + 6 = 0$$
$$\%1 := \frac{1}{2}\arctan\!\left(\frac{3}{5}\right)$$

Complete the square in X and Y by

- q12 := completesquare(q11, [X,Y]);

$$q12 := \left(-\frac{1}{2}+\frac{1}{2}\sqrt{34}\right)\left(Y-\frac{3\sin(\%1)+8\cos(\%1)}{-1+\sqrt{34}}\right)^2$$

$$-\frac{\left(-\frac{1}{2}+\frac{1}{2}\sqrt{34}\right)\left(9\sin(\%1)^2+48\cos(\%1)\sin(\%1)+64\cos(\%1)^2\right)}{(-1+\sqrt{34})^2}$$

$$+\left(-\frac{1}{2}\sqrt{34}-\frac{1}{2}\right)\left(X-\frac{-3\cos(\%1)+8\sin(\%1)}{\sqrt{34}+1}\right)^2$$

$$-\frac{\left(-\frac{1}{2}\sqrt{34}-\frac{1}{2}\right)\left(9\cos(\%1)^2-48\cos(\%1)\sin(\%1)+64\sin(\%1)^2\right)}{(\sqrt{34}+1)^2}+6=0$$

$$\%1 := \frac{1}{2}\arctan\left(\frac{3}{5}\right)$$

Since the expressions for the parameters involved have gotten cumbersome, change them all to floating-point form by

- q13 := evalf(q12);

$$q13 := 2.415475948\,(Y-1.761665269)^2-1.454545455$$
$$-3.415475948\,(X+.1106253039)^2=0$$

The center of the hyperbola is at (-.1106253039, 1.761665269). Further progress to standard form is possible by the following steps, similar to those performed in Example 1.

- q14 := q13 + (1.454545455 = 1.454545455);

$$q14 := 2.415475948\,(Y-1.761665269)^2-3.415475948\,(X+.1106253039)^2=$$
$$1.454545455$$

- q15 := q14/1.454545455;

$$q15 := 1.660639714\,(Y-1.761665269)^2-2.348139714\,(X+.1106253039)^2=$$
$$1.000000000$$

The coefficients of X^2 and Y^2 are useful. Hence,

- c1 := coeff(expand(lhs(q15)),X,2);
  c2 := coeff(expand(lhs(q15)),Y,2);

$$c1 := -2.348139714$$

$$c2 := 1.660639714$$

If the standard form of the hyperbola is taken as $[(Y - k)/a]^2 - [(X - h)/b]^2 = 1$, then a and b are given by $1/\sqrt{(c2)}$ and $1/\sqrt{(|c1|)}$, respectively. We leave the reader the opportunity to obtain the equations of the asymptotes and the foci.

Exercises

1. Find an equation of:

    (a) the circle that passes through the points (1, 1), (3, 2), and (0, 6).

    (b) the ellipse with center at (0, 0) which passes through the points (3, 10), (-3, 10), (1, -16), and (-1, -16).

    (c) a hyperbola that passes through the points (3, 10), (3, -10), (1, -16), and (-1, -16). Can you find a second hyperbola that passes through these four points?

2. Suppose that a hyperbola has its major axis parallel to the x-axis, its center on the y-axis, and passes through the points (4, 2) and (5, 1). Find an equation of such a hyperbola. Does more than one hyperbola satisfy these conditions?

3. Find the coordinates of the vertex and focus, and plot

    (a) $x^2 - 4x + 64 = 12y$ (c) $x^2 - 4xy + 4y^2 - 8\sqrt{5}x - 4\sqrt{5}y = 0.$

    (b) $y^2 - 4y + 64 = 12x$

4. Find the coordinates of the center and foci, and plot

    (a) $4x^2 + 9y^2 - 24x - 36y + 36 = 0$

    (b) $4x^2 - 9y^2 - 24x + 36y - 36 = 0$

    (c) $73x^2 - 72xy + 52y^2 + 380x - 160y + 400 = 0.$

5. In the x_1, y_1 system, a hyperbola is defined by $\dfrac{y_1^2}{4} - \dfrac{x_1^2}{4} = 1$. The origin of the x_1, y_1 system is at the point (5, 2) in the x,y system, and the positive x_1-axis makes a positive angle of $25°$ with the positive x-axis. Write the equation of the hyperbola in terms of x and y. Plot the hyperbola.

6. Given the two points P and Q, take the distance between them as 2a, and find the locus of a point such that the sum of the squares of its distances from P and Q is equal to $4a^2$.

    (a) P = (-2, 4), Q = (2, 4)   (b) P = (1, 1), Q = (5, 9).

7. Find the point on the parabola which is closest to the focus of the parabola.

   (a) $y^2 = 4x - 8$ **(b)** $x^2 - 12x + 4y + 20 = 0$.

8. The comet Zagon follows a parabolic orbit, with the planet Zyfor at the focus of the parabola. Zagon is 95 million miles from Zyfor when the line from Zyfor to the comet is perpendicular to the axis of the parabola. How close does the comet come to the planet?

9.   Make a movie that shows the comet of Exercise 8 as it moves around the planet Zyfor.

10. How close does the comet in Exercise 8 come to the planet if the comet is 100 million miles from Zyfor when the line from the planet to the comet first makes an angle of 45° with the axis of the parabola?

11. Find the area inside of

   (a) the ellipse $4x^2 + 8y^2 - 40x - 48y + 140 = 0$

   (b) the conic $5x^2 + 4xy + 5y^2 = 9$.

12. In the same window, plot $y^2 = 6 - x$ and $y^2 = 4x + 16$. Find the acute angle between the curves at their point of intersection.

13. Plot the conic and find the point on the graph which is closest to the origin.

   (a) $4x^2 + 8y^2 - 40x - 48y + 140 = 0$

   (b) $1.75x^2 - 2.60xy + 3.25y^2 + 5.07x - 24.78y + 48 = 0$.

Projects

1. Find an equation of the conic which passes through the given points. Then plot the conic, find the coordinates of the foci, and the ends of the major axes. (Can you plot the points by hand first and predict what type of conic passes through the points?)

   (a) (2, 0), (-2, 0), (0, 2), (-1, -1), (4, 2)

   (b) (1, 0), (2, 2), (6.5, 6), (7, 2.5), (10, 2).

2. Find an equation of a parabola that passes through the points (1, 3), (2, 7), and (3, 4). Is your parabola the only one which passes through these points? Find a point on the parabola which is closest to the focus.

3. Assume that a headlight is formed by revolving the parabola $y = 4x^2$ about its axis, and the bulb is located at the focus. Prove that any ray of light from the bulb, after striking the surface of the paraboloid, travels parallel to the axis.

    Suggestion: Draw line BC parallel to the axis of the parabola, and line AD tangent to the parabola at point B; then show that $\angle ABF = \angle DBC$ by showing that $\overline{AF} = \overline{BF}$ in $\triangle ABF$.

4. (a) Show that if the conic $A x^2 + B x y + C y^2 + D x + E y + F = 0$ is transformed to $A'xp^2 + B'xp\, yp + C'yp^2 + D'xp + E'yp + F' = 0$ by rotation through any angle θ, then $B^2 - 4 A C = (B')^2 - 4 A'B'$.

    (b) Show that if θ is chosen to make $B' = 0$, a simple test can be stated in terms of the sign of $A'C'$ which tells if the conic is a parabola, ellipse, or hyperbola. Now use part (a) to state the test in terms of $B^2 - 4 A C$.

5. Use the final result from Exercise 4, part (b), to classify each conic:

    (a) $12 x^2 + 8 x y - 9 y^2 + 64 x + 30 y = 0$

    (b) $16 x^2 + 24 x y + 9 y^2 - 85 x - 30 y + 175 = 0$

    (c) $4 x^2 - 4 x y + 5 y^2 - 10 x + 10 y - 10 = 0$.

6. In a faraway galaxy, three planets follow elliptical orbits with the center of the star SX10 at one focus (the "rightmost") of the orbit of planet B. Assume that the orbits of the planets have a common center, as well as a common major axis; call it the x-axis. For the given semi-major axis length a and eccentricity e, plot the three orbits in a common window.

Planet	a (* 10^8 km)	e
A	1.0	0.14
B	1.3	0.28
C	3.2	0.56

7. Refer to Exercise 6.

    (a) What size diameter of SX10 would put both foci of an orbit inside the star? For which planet?

    (b) Suppose the major axis of planet C is tilted $30°$ counterclockwise. Plot the three orbits in a common window.

Chapter 13

Polar Coordinates and Parametric Equations

New Maple Commands for Chapter 13

map(function,target);	applies *function* to each object in target
plot([x(t),y(t),t=a..b]);	plot of parametric curve x = x(t), y = y(t)
plot([r(t),t,t=a..b],coords=polar);	polar plot of r = r(t)

Introduction

Polar coordinates offer an alternative description of the xy-plane. This description corresponds to a view of the world through a radar screen. As the radar antenna rotates through a full circle, sending out its microwave beam, objects detected are reported back to the screen in terms of the angle through which the beam has rotated, and the distance from the antenna. While radar screens usually measure angles counterclockwise from the line "straight up," polar coordinates use the line "going to the right" to measure the zero angle.

Maple treats polar coordinates as a case of parametric equations. Hence, this chapter includes a study of representing curves via functions describing each coordinate separately. The image here is that of an object moving in the plane, say, and having two range finders, one to measure displacements in the horizontal direction and the other, in the vertical. The path of the motion can be reconstructed from the data, provided the instants, that is, the times, when data points were recorded are synchronized. The resulting table of values for t, x, and y giving functions x = x(t) and y = y(t) is the prototype for thinking about parametric equations.

Examples

Example 1

Given the polar curves r1 = 3 - 3 sin(t) and r2 = 2 - 3 sin(t), find the area of the region inside of r1 and outside r2.

- r1 := 3 - 3*sin(t);
  r2 := 2 - 3*sin(t);

$$r1 := 3 - 3\sin(t)$$

$$r2 := 2 - 3\sin(t)$$

Assign the plot data-structures for the graphs of these curves to variables p1 and p2, thickening the curve for the first.

- p1 := plot([r1, t, t = 0..2*Pi], coords = polar, thickness = 3):
  p2 := plot([r2, t, t = 0..2*Pi], coords = polar):

View the graphs by entering the variable names p1 and p2. The colon (:) at the end of each command above is essential for suppressing the print-out of the actual data structure itself.

• p1;

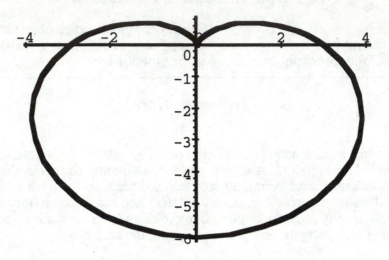

• p2;

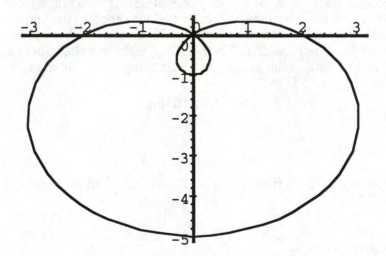

Using the display command from the plots package, superimpose the two plot data-structures p1 and p2. Note that instead of thickening one curve, we could have used a different line pattern, or even used different colors on a color screen. The thickened curve used here seemed to be the most visible for our present purposes.

• with(plots):

• display([p1,p2]);

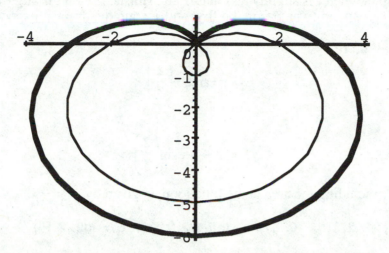

Thus, r2 generates the thin graph with the little loop, and also forms the inner boundary. However, we don't know the values of t for which the little loop is traced. We next define a function p(T) that returns a polar graph of r2 for $0 \le t \le T$. Experiments with this function, and even an animation, will give insight into the tracing of r2.

- p := T->plot([r2, t, t = 0..T],coords = polar);

$$p := T \to \text{plot}([r2, t, t = 0 .. T], coords = polar)$$

- display([seq(p(Pi/10*k),k=0..20)],insequence = true);

The reader is encouraged to experience this animation live and to experiment with calls to p(t). An alternative strategy is based on computing where r2 is zero, corresponding to the passage of the curve through the origin. Begin with a plot of r2 against t.
- plot(r2,t=0..2*Pi);

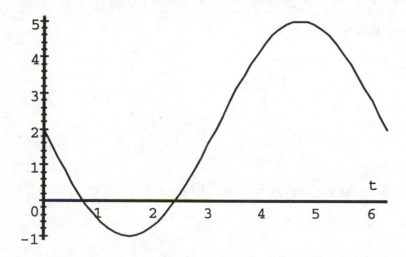

The little loop begins when t is slightly less than 1 and finishes when t is slightly greater than 2. These values correspond to angles in the first and second quadrants, respectively.

• q1 := solve(r2 = 0, t);

$$q1 := \arcsin\left(\frac{2}{3}\right)$$

• q2 := Pi - q1;

$$q2 := \pi - \arcsin\left(\frac{2}{3}\right)$$

The area inside r2, including the area in the little loop, is given by

• a2 := Int((1/2)*r2^2, t = 0..q1) + Int((1/2)*r2^2, t = q2..2*Pi);

$$a2 := \int_0^{\arcsin\left(\frac{2}{3}\right)} \frac{1}{2}(2 - 3\sin(t))^2\, dt + \int_{\pi - \arcsin\left(\frac{2}{3}\right)}^{2\pi} \frac{1}{2}(2 - 3\sin(t))^2\, dt$$

If we had integrated on $0 \le t \le 2\pi$ we would have gotten the area of the little loop twice, once as the angle t traversed the interval [q1, q2] and again as t traversed the interval $[\pi, 2\pi]$. Remember, the element of area in polar coordinates is the ray extending from the origin to the bounding curve. This ray perforce sweeps over the little loop as it extends out to the larger boundary of r2. So if the ray is allowed to sweep out the little loop by integrating over $q1 \le t \le q2$, the area of the little loop would be computed twice.

The area inside r1 is given by

• a1 := Int((1/2)*r1^2, t = 0..2*Pi);

$$a1 := \int_0^{2\pi} \frac{1}{2}(3 - 3\sin(t))^2\, dt$$

The area inside r1 but outside r2 is given by the difference

• q3 := value(a1 - a2);

$$q3 := \frac{37}{4}\pi - \frac{17}{2}\arcsin\left(\frac{2}{3}\right) - 3\sqrt{5}$$

• evalf(q3);

$$16.14884304$$

Example 2

If $x(t) = 4 - t^2$ and $y(t) = t^2 - 2t - 2$ parametrically define a curve, obtain its graph, noting that $x(t)$ is increasing when $t < 0$ and that $y(t)$ is increasing when $t > 1$. Can you guess the general shape of the graph from this information?

In naming the expressions for $x(t)$ and $y(t)$, keep in mind the rule that one should never use for a name on the left of an assignment operator (:=) a "working" variable that might be needed on the right. Hence, we use xt and yt rather than x and y.

• xt := 4 - t^2;
  yt := t^2 - 2*t - 2;

$$xt := 4 - t^2$$

$$yt := t^2 - 2t - 2$$

• plot([xt, yt, t = -2..2]);

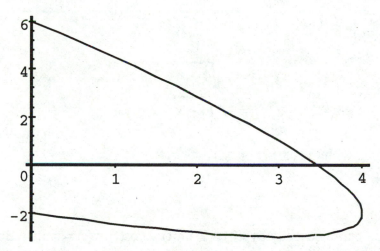

Example 3

For the curve in Example 2, express y as y(x). This process is also called "eliminating the parameter."

The most direct way to have Maple perform this elimination is via

- q4 := solve({xt = x, yt = y}, {y, t});

$$q4 := \{\, y = -2\,\text{RootOf}(-4 + \_Z^2 + x) + 2 - x, t = \text{RootOf}(-4 + \_Z^2 + x)\,\}$$

- allvalues(q4[1]);

$$y = -2\sqrt{4-x} + 2 - x, y = 2\sqrt{4-x} + 2 - x$$

The rule about not using either x or y as names for x(t) and y(t) was essential in setting up the equations that we solved in q4. However, if this solution seems like too much of a "black box," then direct Maple to solve the first equation (xt = x) for t = t(x) and substitute this into the second to obtain y = y(x). Note how we fastidiously avoid using t as the name of this solution since t is still a "working" variable in use on the right.

- q5 := solve(xt = x, t);

$$q5 := -\sqrt{4-x}, \sqrt{4-x}$$

The substitution step must be done separately for each branch. To make these solutions look like the ones Maple generated earlier, we will create equations with y on the left.

- y = subs(t = q5[1], yt);
  y = subs(t = q5[2], yt);

$$y = 2\sqrt{4-x} + 2 - x$$
$$y = -2\sqrt{4-x} + 2 - x$$

These solutions agree with the earlier ones.

Example 4

Plot the curve given parametrically by $x(t) = 2\sin(t) + \cos(t)$, $y(t) = \sin(t)$. Before plotting, predict the general shape of the curve by eliminating the variable t and computing $B^2 - 4AC$ for the resulting conic.

The astute reader will understand that not every parametrically given curve with formulas containing trigonometric functions is automatically a curve in polar coordinates. This curve, parametrically given, is in a cartesian frame.

- xt := 2*sin(t) + cos(t);
  yt := sin(t);

$$xt := 2\,\sin(t) + \cos(t)$$

$$yt := \sin(t)$$

We adopt the strategy of solving the second equation for $t = t(y)$ and substituting this into the first, to obtain $x = x(y)$. From this, we will obtain the quadratic $Ax^2 + Bxy + Cy^2 + Dx + Ey + F = 0$.

- q6 := solve(yt = y, t);

$$q6 := \arcsin(y)$$

- q7 := x = simplify(subs(t = q6, xt));

$$q7 := x = 2\,y + \sqrt{1 - y^2}$$

To eliminate the radical we will have to **isolate** it, then square both sides of the resulting equation. We square both sides of an equation by mapping the squaring function onto (both sides of) the equation.

- readlib(isolate):

- q8 := isolate(q7, sqrt(1 - y^2));

$$q8 := \sqrt{1 - y^2} = x - 2\,y$$

- q9 := map(x->x^2, q8);

$$q9 := 1 - y^2 = (x - 2\,y)^2$$

The general form of the quadratic has all terms on the left side of the equation.

- q10 := lhs(q9) - expand(rhs(q9));

$$q10 := 1 - 5\,y^2 - x^2 + 4\,y\,x$$

If the discriminant is negative the quadratic is an ellipse.

- 4^2   -   4*(-1)*(-5);

-4

• plot([xt, yt, t = 0..2*Pi]);

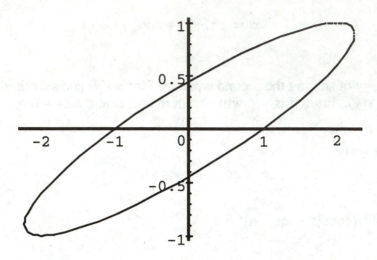

Example 5

Parametrize the Trisectrix, $x^3 + xy^2 + y^2 - 3\,x^2 = 0$, by intersecting the curve with the line $y = mx$. This will yield, for example, $x = x(m)$ from which follows $y = y(m)$.

Begin with an implicit plot of the Trisectrix by using the **implicitplot** command in the plots package.

• q11 := x^3 + x*y^2 + y^2 - 3*x^2 = 0;

$$q11 := x^3 + xy^2 + y^2 - 3\,x^2 = 0$$

• implicitplot(q11, x = -5..5, y = -5..5, numpoints = 1000);

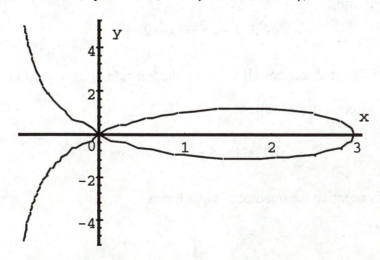

Substitute y = mx into the equation of the Trisectrix.

• q12 := subs(y = m*x, q11);

$$q12 := x^3 + x^3 m^2 + m^2 x^2 - 3 x^2 = 0$$

Solve q12 for x = x(m).

• q13 := solve(q12, x);

$$q13 := 0, 0, -\frac{m^2 - 3}{1 + m^2}$$

Clearly, we want the third solution, which then defines x = x(m). Since y = mx, we also obtain y = y(m).

• xm := q13[3];
  ym := m*xm;

$$xm := -\frac{m^2 - 3}{1 + m^2}$$

$$ym := -\frac{m(m^2 - 3)}{1 + m^2}$$

• plot([xm, ym, m = -5..5]);

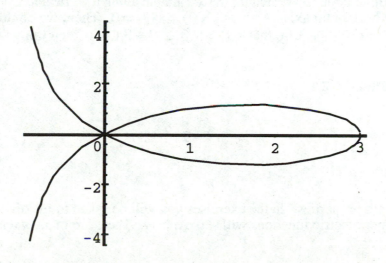

Example 6

Find a parametric representation $x = x(t)$, $y = y(t)$, $0 \le t \le \pi$, for the upper half of the ellipse $x^2/25 + y^2/4 = 1$.

Since $x(t) = \cos(t)$, $y(t) = \sin(t)$, $0 \le t \le$ Pi, is the parametric representation of the top half of a circle that is traced out counterclockwise, we seek to convert the problem with the ellipse into the one for the circle. If we define $u = x/5$ and $v = y/2$, the ellipse is then of the form $u^2 + v^2 = 1$. Since $u = \cos(t)$ and $v = \sin(t)$ would trace a circle, we have $x = 5 \cos(t)$ and $y = 2 \sin(t)$ as a parametric representation of the ellipse.

- plot([5*cos(t), 2*sin(t), t = 0..Pi], scaling = constrained);

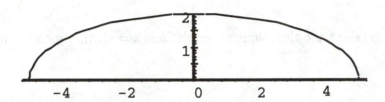

To trace this ellipse parametrically in a clockwise sense, reverse the direction of the parameter by changing t to π - s, $0 \le s \le \pi$. To trace the upper half of this ellipse in a clockwise sense for a parameter z, $\pi \le z \le 2\pi$, replace t with z - π.

Example 7

Define a curve parametrically so that it has the shape of an upper-case S. Note that y will always be increasing, but x will increase, then decrease, and then increase again. You might find this easier to think about by sketching the x-variable along a vertical axis with the t-variable along the horizontal axis. Also, say $x'(t) = x'(3) = 0$. Then, we could take $x'(t) = (t - 1)(t - 3) = t^2 - 4t + 3$, so that $x(t) = t^3/3 - 2t^2 + 3t$. To keep things simple, let's try $y(t) = t$.

- xt := t^3/3 - 2*t^2 + 3*t;
  yt := t;

$$xt := \frac{1}{3}t^3 - 2t^2 + 3t$$

$$yt := t$$

This has the required properties. In the exercises you will be asked to make some "loops." Polynomials or trigonometric functions will serve nicely. There are many ways to create these curves.

Exercises

1. Find a few points on each curve, check it for symmetry and other helpful properties (such as rectangular coordinates), and make a sketch; then use Maple to plot the curve.

   (a) $r = 5 \sin 3\theta$ **(g)** $r^2 = 9 \cos 2\theta$

   (b) $r = 5 \cos 2\theta$ **(h)** $r = \dfrac{4}{1 - \cos \theta}$

   (c) $r = 2 - 2 \cos \theta$ **(i)** $r = \sec \theta$

   (d) $r = 2 - 3 \cos \theta$ **(j)** $r = -2 \sin \theta$

   (e) $r = \dfrac{\theta}{2} \; (-2\pi \le \theta \le 2\pi)$ **(k)** $r = 1 + \sin \dfrac{\theta}{2}$

   (f) $r = \exp\left(\dfrac{\theta}{2}\right)$ **(l)** $r = 1 + 2 \sin \dfrac{\theta}{2}$.

2. Make a 33-frame movie of the point $[2 - 3 \cos \theta, \theta]$ as θ takes on values from 0 to 3.2. (Polar coordinates.)

3. Plot the curves and find <u>all</u> intersection points. (Naive use of **fsolve** or **solve** will not do the job here.)

   (a) $r_1 = 1 + \cos \theta$, $r_2 = 1 - \sin \theta$ **(b)** $r_1 = 2 \sin \theta$, $r_2 = 1 + \cos \theta$.

4. Find the area of the region which is inside of r_1 and outside of r_2.

   (a) $r_1 = 2 \sin \theta$, $r_2 = 1 + \cos \theta$

   (b) $r_1 = 2 - 2 \sin \theta$, $r_2 = 2 \sin 2\theta$

   (c) $r_1 = 2 - 2 \sin \theta$, $r_2 = 3 \sin 2\theta$

   (d) $r_1 = 2 - 2 \sin \theta$, $r_2^2 = 16 \sin \theta$. (Some of the intersection points are elusive here.)

5. Plot $r = \dfrac{4}{1 - \sin \theta}$. Note that the point $P = (-4, \pi)$ lies on the graph. Do the coordinates of P satisfy the equation? Can you show algebraically that P is on the graph?

6. Plot $r = \dfrac{d}{\cos(\theta - \alpha)}$ for

(a) $d = 2$, $\alpha = \dfrac{\pi}{6}$ (b) $d = 2$, $\alpha = \dfrac{\pi}{3}$ (c) $d = 4$, $\alpha = \dfrac{\pi}{3}$

(d) Plot parts (b) and (c) on the same coordinate system.

7. Obtain parametric plots of each of the following. Then, if possible, eliminate the parameter and plot the resulting explicit or implicit representation. Explain any discrepancies in the plots.

(a) $x = 1 + 2\cos t$, $y = 2 + 4\sin t$

(b) $x = 2 + \dfrac{1}{t}$, $y = 1 - \dfrac{1}{t}$

(c) $x = 8\cos^2 t$, $y = 8\sin^2 t$

(d) $x = 8\cos^3 t$, $y = 8\sin^3 t$

(e) $x = 2\sin t + \cos t$, $y = 2\cos t - \sin t$

(f) $x = 2\sin t + \cos t$, $y = 2\cos t$

(g) $x = \sin t$, $y = 2\sin 5t$, $0 \le t \le 2\pi$

(h) $x = t^3 - 2t$, $y = t^3 - 2t^2$, $-2 \le t \le 3$

(i) $x = t^2 - t$, $y = 2t^2 - t^4$, $-1 \le t \le 1$

(j) $x = 2\cos t + \cos 2t$, $y = 2\sin t - \sin 2t$.

8. Make a 21-frame movie of the point $(t^2 - t, 2t^2 - t^4)$ as t takes on values from -1 to 1. (Parametric form.)

9. Parametrize each equation by finding the intersection with the line $y = mx$ and then plot. (Some of these may be more familiar in rectangular form.)

(a) $y^2 = 8x$

(b) $x^3 + y^3 - 4xy = 0$ (The Folium of Descartes)

(c) $x^3 + xy^2 + x^2 - y^2 = 0$ (The Strophoid)

(d) $(x^2 + y^2)^2 = 4\,x^2 y$ (The Bifolium).

10. Find a parametric representation of the curve which is specified and then plot the parametric representation found.

(a) A circle of radius 1, traversed in a counterclockwise direction starting at the rightmost point.

(b) A circle of radius 1, traversed in a clockwise direction starting at the rightmost point.

(c) A circle of radius 4, traversed in a clockwise direction starting at the leftmost point.

(d) A circle of radius 4, traversed in a counterclockwise direction starting at the top.

(e) An ellipse, major axis horizontal and length 10, minor axis of length 4, starting at the rightmost point.

(f) An ellipse, major axis vertical and of length 8, minor axis of length 6, starting at the leftmost point.

11. Find a parametric representation of a curve which has the general shape shown. Then plot your curve.

(a)   (b)

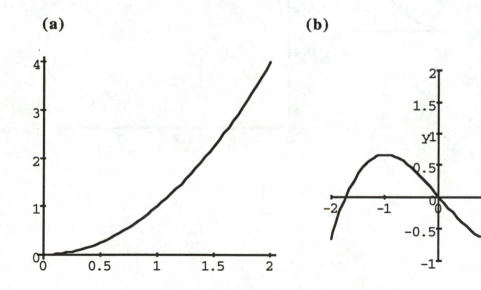

(c)

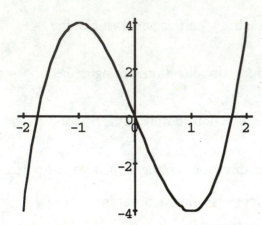

(d)

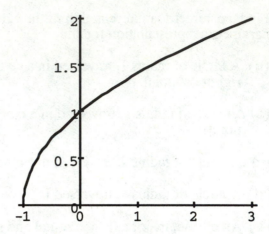

(e)

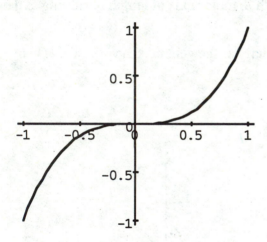

(f)

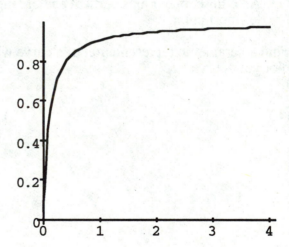

(g)

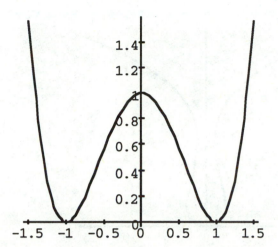

(h)

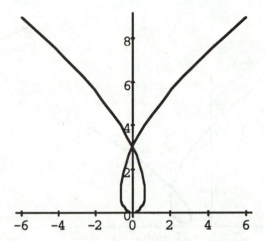

(i)

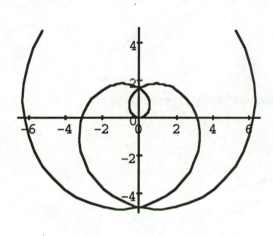

(j)

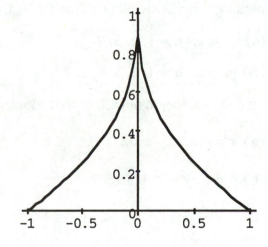

(k)   **(l)**

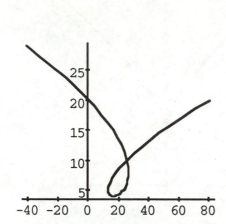 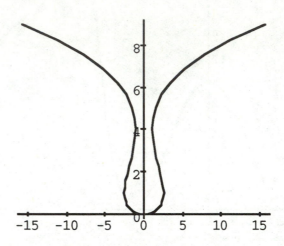

12. Lissajous curves (named after the nineteenth century French physicist Jules Antoine Lissajous) have the parametric form $x = A \cos (mt + \alpha)$, $y = B \sin (nt + \beta)$. Taking A, B, α, and β as 1, plot Lissajous curves for the following values of m and n. Try to predict the appearance of some of these curves. Can you use **for** loops to automate this plotting? Try some other values for A, B, α, and β.

   (a) $m = 1$, $n = 1, 2, 3, 4$ **(c)** $m = 3$, $n = 3, 4, 5, 6$.

   (b) $m = 2$, $n = 1, 2, 3, 4, 5, 6$

13. Find the arc length of each of the following parametrically given curves.

   (a) $x = \cos^3 t$, $y = \sin^3 t$, $0 \le t \le \pi$

   (b) $x = t^3$, $y = t^2$, $0 \le t \le 2$

   (c) $x = e^{-t} \cos t$, $y = e^{-t} \sin t$, $0 \le t \le \pi$.

14. Find an equation of the line tangent to the curve $x = e^t$, $y = e^{-t} + 1$, where $t = 0$.

15. Find a point on the curve $x = t^2 - t$, $y = 2 t^2 - t^4$, $-1 \le t \le 1$, where the tangent line

   (a) has slope 1   **(b)** has slope -1   **(c)** is vertical.

16. Find the point at which the slope of the tangent line is maximum on $x = t^2 - t$, $y = 2 t^2 - t^4$, $-1 \le t \le 1$.

Projects

1.   Plot, in polar coordinates, the graph of $r = 1 - \sin 5\theta$. Find the values of θ for which $r = 0$. Find the area of the region inside the graph.

2.   Plot, in polar coordinates, the graph of $r = 1 - 4 \sin 5\theta$. Find the values of θ for which $r = 0$. Find the area of the region between the inner and outer branches of the curve. How different is the problem if you work with $r = 1 - 3 \sin 5\theta$?

3.   Plot the path of a pebble caught in the treads of a bicycle tire that rolls along a level road. Let the radius of the wheel be a, and describe the position of the pebble using parametric equations. Imagine a line from the center of the wheel to the pebble. The angle through which this line turns is an effective parameter. (The path is a cycloid.)

4.   Plot the path of the pebble on the wheel in Problem 3 if the wheel rolls around the inside of a larger wheel of radius b. Assume that the bicycle wheel has radius a.

5.   A hula hoop of 40-inch inner circumference makes one complete cycle around a girl with a 20-inch waist, which has negligible motion. What is the length of the circuitous path of a given point on the hoop as it travels from a contact point back to the same point?
                                            - Litton Industries, Problematical Recreations (The Bent)

6.   For each curve below, find all values of t for which the given curve intersects itself.

   (a) $x = t^3 - 1.5\, t^2 - 6\, t,\quad y = t^2 - 2t$.

   Can you find the finite area determined here?

   (b) $x = \frac{1}{5}t^5 - t^4 - \frac{7}{3}t^3 + 11\, t^2 + 24\, t,\quad y = \frac{1}{6}t^6 - 1.18\, t^5 + 0.9\, t^4 + 6.5\, t^3 - 4.7\, t^2 - 8.8\, t.$

   (Be careful on this one; don't miss any points.)

7.   Typesetting began with Johannes Gutenberg in the mid-1400s and became mechanized when a steam-powered press was built in London early in the nineteenth century, followed by Ottmar Mergenthaler's Linotype machine nearer the end of that century. Around 1980, Donald Knuth developed Metafont, a computer language for creating fonts, which allowed the user to modify fonts by assigning different parameter values. By 1990, Summer Stone (an Adobe employee whose education included calligraphy classes and a mathematics degree) had created fonts such as Serif and Sans. Today, hundreds of fonts are available to personal computer users. (An interesting series of articles on "Type" appeared in Mac World, July 1991 through October 1991.)

   This project involves cubic curves in the plane which can be used to form letters, or more general characters and shapes. Curves defined in the usual rectangular form,

$y = f(x)$, do not work well for this kind of application, and parametric curves are used here.

(a) Define a parametric curve using cubics $x(t)$, $y(t)$, for $0 \leq t \leq 1$, such that the curve $(x(t), y(t))$ passes through the points $p_1 = (0, 2)$, $p_2 = (3, 0)$, $p_3 = (7, 3)$ and

$p_4 = (4, 5)$. More specifically, make $(x(0), y(0)) = p_1$, $(x(\frac{1}{3}), y(\frac{1}{3})) = p_2$,

$(x(\frac{2}{3}), y(\frac{2}{3})) = p_3$ and $(x(1), y(1)) = p_4$. Plot your curve.

(b) Use one (or more) curve(s) similar to your result in part (a) to form some of the letters C, O, S, U, d, g, p, k. (On a sheet of graph paper, draw the letter and then pick 4 or 7 or 10 or ... nodes.)

(c) Use the Bezier* curve as defined by $(xb(t), yb(t))$ to form the letters given in part (b). Your four specified nodes $(x(k + 1), y(k + 1))$ for $k = 0..3$ will affect the shape of a section of each letter.

$$xb(t) = \sum_{k=0}^{3} \binom{3}{k} x(k+1)(1 - t)^{3-k} t^k, \qquad yb(t) = \sum_{k=0}^{3} \binom{3}{k} y(k+1)(1 - t)^{3-k} t^k$$

Note that although both $xb(t)$ and $yb(t)$ are cubic in t, and should be similar to your result in part (a), they generate slightly different curves.

(d) Compare the letters "S" and "p", as generated by your curves in parts (b) and (c). Do the functions pass through all of the nodes? Through how many? How can you change the shape of a letter? How would you make it wider, or how would you make the descender longer? Do you see why rectangular curves of the form $y = f(x)$ are not good candidates for these problems?

(e) Do any of your letters make an abrupt change of direction at a connecting node? (A connecting node is one common to two sections.) Can you think of some ways to smooth a curve at such a point? (Splines can be used to eliminate the "jerkiness" that occurs at some of the connecting nodes. The Maple help files contain information and examples for **spline** and **bspline**. The next project will introduce splines.

*Named for the French mathematician, Pierre Bezier. Bezier curves and surfaces are supported by software such as AutoCad and CorelDraw.

8. (a) Define a parametric curve using cubics $x(t)$, $y(t)$, such that the curve $(x(t), y(t))$ starts at the point $(3, 7)$, passes through $(4, 3)$ and $(3, 1)$, and stops at $(2, 3)$. Define a similar curve that starts at $(2, 3)$, passes through $(3, 8)$ and $(6, 7)$, and stops at $(3, 5)$. One method for doing this is outlined in Problem 7, part (a).

(b) Display these two curves in the same window. You should see a "p" (or maybe part of a "Snoopy"). Can you see a section of the figure change direction at a point?

(**c**) Fix this lack of smoothness at this point. (The point is the connecting point of the two curves.) One method of solution is outlined here:

Define a parametric curve using quadratics x(t), y(t), such that the curve (x(t), y(t)) starts at (3, 7) and ends at (4, 3). Define another such curve that starts at (4, 3) and ends at (3, 1). In addition, make the slopes of these curves agree at the connection point. Continue this approach to the last point. When you display the curves in the same window, you should see a continuous, smooth curve.

(**d**) Repeat parts (**a**), (**b**), and (**c**) for some letter or shape of your choice.

9.   In this project you are asked to generate polynomial approximations to several polar functions. Note that in these examples the Taylor polynomial gives a more computable form of the function.

Find a Taylor polynomial of degree n which gives a good approximation to $r = f(\theta)$ for θ near θ_0. Plot, in polar coordinates, $r = f(\theta)$ and the Taylor approximation to $r = f(\theta)$ in the same window.

(**a**) $r = 1 - \cos\theta$, $n = 4, \theta_0 = 0$

(**b**) $r = 2\sin 3\theta$, $n = 4, \theta_0 = 0$

(**c**) $r = \cos 2\theta$, $n = ?$, $\theta_0 = \dfrac{\pi}{8}$

(**d**) $r = 3 - 4\cos\theta$, $n = ?$, $\theta_0 = ?$

(**e**) $r = \dfrac{1}{\theta}$, $n = ?$, $\theta_0 = \dfrac{\pi}{2}$.

For more on this topic, see "Taylor Polynomial Approximations in Polar Coordinates" by Shelly Gordon, *The College Mathematics Journal,* Vol. 24, no. 4, September 1993.

Chapter 14

Vectors and Three Dimensional Space

New Maple Commands for Chapter 14

plot3d(f(x,y),x=a..b,y=c..d);	plots 3-D surface z = f(x,y)
with(linalg);	load the linear algebra package
crossprod(v1,v2);	compute cross product of vectors **v1** and **v2**
det(A);	compute determinant of matrix A
dotprod(v1,v2);	compute dot product of vectors **v1** and **v2**
normalize(v);	return unit vector **v** / (length of **v**)
stack(v1,v2);	form matrix whose rows are vectors **v1**, **v2**
vector([a,b,c]);	creates vector a **i** + b **j** + c **k**
with(plots);	loads plots package
cylinderplot(r(t,z),t=a..b,z=c..d);	plots r(t,z) in cylindrical coordinates
spacecurve([x(t),y(t),z(t)],t=a..b);	plots 3-D curve x = x(t), y = y(t), z = z(t)
sphereplot(r(t,f),t=a..b,f=c..d);	plots 3-D surface r(t,f) in sperical coordinates
with(student);	loads student package
completesquare(quad,[x,y,z]);	completes square in quadratic in x, y, z

---

Introduction

Multivariate calculus generally starts with a discussion of lines and planes in three-dimensional xyz-space, the space in which we normally seem to live. In fact, the philosopher Kant claimed that our ability to perceive the world in three dimensions was a "categorical imperative," or a built-in faculty characterizing human intelligence. Whatever we might think of Kant's philosophy, we cannot escape this facet of our experiences: that there is length, width, and height to our observations. Moreover, in the book "Flatland," E. Abbot grapples with the problem of a two-dimensional being struggling to comprehend an essentially unknowable three-dimensional world. So, mathematicians have not been the only ones to observe that our imaginations have some ability to present three-dimensional representations to our minds. This chapter is predicated on the conviction that with experience, practice, and training, we can improve what skills we do have to think about three-dimensional objects that have only an immaterial presence in our minds.

Examples

Example 1

The quadric surface defined by $x^2 + y^2 + z^2 - 8x - 6y + 4z + 25 = 0$ is a sphere, and by completing the square the center can be found. The point on the sphere closest to the origin is then on the line connecting the center with the origin.

- q := x^2 + y^2 + z^2 - 8*x - 6*y + 4*z + 25 = 0;

$$q := x^2 + y^2 + z^2 - 8x - 6y + 4z + 25 = 0$$

- with(student):
- q1 := completesquare(q, [x, y, z]);

$$q1 := (z+2)^2 - 4 + (y-3)^2 + (x-4)^2 = 0$$

The center is at (4, 3, -2), and the line connecting the center with the origin is given parametrically by x(t) = 4t, y(t) = 3t, z(t) = -2t. Making these substitutions into the implicit function in q yields a single equation in t.

- q2 := subs(x = 4*t, y = 3*t, z = -2*t, q);

$$q2 := 29\, t^2 - 58\, t + 25 = 0$$

- q3 := solve(q2, t);

$$q3 := 1 + \frac{2}{29}\sqrt{29},\, 1 - \frac{2}{29}\sqrt{29}$$

As t varies from 0 to 1, the point (x(t), y(t), z(t)) on the line connecting the origin to the center of the sphere moves from the origin to the center of the sphere. Thus, t is a measure of distance along the line from the origin to the sphere, and the second value of t in q3 yields the shortest distance.

- T := q3[2];

$$T := 1 - \frac{2}{29}\sqrt{29}$$

- sqrt(expand((4*T)^2 + (3*T)^2 +(-2*T)^2));

$$\sqrt{29} - 2$$

Example 2

Write the vector **v** = [3, -2, 0]^t as a linear combination of the column vectors **v1** = [1, 2, 0]^t, **v2** = [1, 0, 2]^t, and **v3** = [4, -1, 1]^t. All four vectors are column vectors since the superscript "t" refers to "transpose."

The vector v is a linear combination of **v1**, **v2**, and **v3** if the equations represented by the identity **v** = a **v1** + b **v2** + c **v3** have solutions for the scalars a, b, and c.

- e1 := 3 = a + b + 4*c;
  e2 := -2 = 2*a -c;
  e3 := 0 = 2*b + c;

$$el := 3 = a + b + 4\,c$$

$$e2 := -2 = 2\,a - c$$

$$e3 := 0 = 2\,b + c$$

- solve({e1, e2, e3}, {a, b, c});

$$\left\{ b = \frac{-1}{2}, a = \frac{-1}{2}, c = 1 \right\}$$

Hence, v = -1/2 **v1** - 1/2 **v2** + **v3**.

Example 3

Find a vector that is orthogonal to each of the vectors **v1** = [2, 2, 1-1]^t and **v2** = [3, 1, -2]^t.

The vector we want is the cross product of **v1** and **v2**, which we can compute in either of two ways, each of which requires the linear algebra package.

- with(linalg):

- v1 := vector([2, 2, -1]);
  v2 := vector([3, 1, -2]);

$$vl := [2 \quad 2 \quad -1]$$

$$v2 := [3 \quad 1 \quad -2]$$

These two vectors are column vectors, even though Maple writes them horizontally!

- crossprod(v1, v2);

$$[-3 \quad 1 \quad -4]$$

This cross product is, again, a column vector. A second way of computing the cross product is by instructing Maple to assemble and compute an appropriate determinant.

- row1 := vector([i, j, k]);

$$rowl := [i \quad j \quad k]$$

- a := stack(row1, v1, v2);

$$a := \begin{bmatrix} i & j & k \\ 2 & 2 & -1 \\ 3 & 1 & -2 \end{bmatrix}$$

• det(a);

$$-3\,i + j - 4\,k$$

Example 4

Find the (minimum) distance between the line L1, given by the equations $x1(s) = 2 + 3s$, $y1(s) = -3 -2s$, $z1(s) = 1 + s$, and L2, the line given by $x2(t) = 5 + 2t$, $y2(t) = -t$, $z2(t) = 4 + t$.

A solution technique based on vectors requires that we find a point P on line L1, a point Q on L2, and the vector **PQ**. In addition, we use the astonishing realization that every pair of skew lines has a common perpendicular. We can find the direction (**v**) of this perpendicular, and the distance between the skew lines is then the length of the projection of **PQ** onto the vector **v**.

By inspection, we can take P as (2, -3, 1) and Q as (5, 0, 4). Then **PQ**, by subtraction, is the vector $[3, 3, 3]^t$. The common perpendicular is the cross product of vectors **v1** and **v2** that give the direction of lines L1 and L2, respectively. Hence,

• v1 := vector([3, -2, 1]);
  v2 := vector([2, -1, 1]);

$$v1 := [3 \quad -2 \quad 1]$$

$$v2 := [2 \quad -1 \quad 1]$$

• v := crossprod(v1, v2);

$$v := [-1 \quad -1 \quad 1]$$

If **r** is the projection of **PQ** along **v**, then the length of **r** is the absolute value of the dot product of **PQ** and **V**, the vector **v** normalized.

• PQ := vector([3,3,3]);
  V := normalize(v);
  dist := abs(dotprod(PQ, V));

$$PQ := [3 \quad 3 \quad 3]$$

$$V := \left[-\frac{1}{3}\sqrt{3} \quad -\frac{1}{3}\sqrt{3} \quad \frac{1}{3}\sqrt{3} \right]$$

$$dist := \sqrt{3}$$

Example 5

Plot the curve given parametrically by the equations x(t) = cos(t), y(t) = sin(t), z(t) = t, where t is in the interval $0 \leq t \leq 6\pi$. Use the Maple **spacecurve** command found in the plots package.

- with(plots):
- spacecurve([cos(t), sin(t), t], t=0..6*Pi, axes = boxed, numpoints = 100);

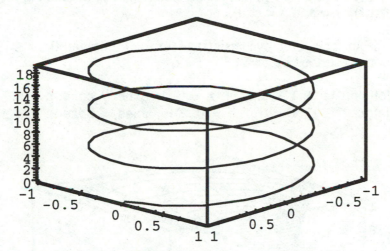

In the plot window, "grab" the 3-D image with the mouse to "rotate" it.

Example 6

Plot the surface z = sin(x) e-y for $0 \leq x \leq \pi$ and $0 \leq y \leq 3$.

- plot3d(sin(x)*exp(-y), x = 0..Pi, y = 0..3, axes = boxed);

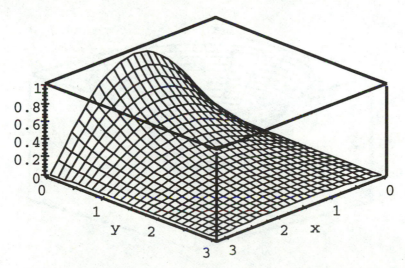

Example 7

If a surface is given in cylindrical coordinates by $r = r(t, z)$ or $z = z(r,t)$, then it can be graphed by Maple's **cylinderplot** command in the plots package. For a surface described by $r = r(t,z)$ the syntax of the **cylinderplot** command is cylinderplot(r, t = a..b, z = c..d). However, if the surface is given by the more natural $z = z(r,t)$ then **cylinderplot** requires the parametric syntax cylinderplot([r(u,v), t(u,v), z(u,v)], u = a..b, v = c..d). Of course, in this latter case, we can identify u with r and t with v so that the surface would be plotted via cylinderplot([r, t, z(r,t)], r = a..b, t = c..d).

For example, plot the surface given by $r = \sin(t)\, z^2$ for $0 \leq t \leq 2\pi$ and $0 \leq z \leq 1$, and then plot the surface of the cone described by $z = 2r$.

- s1 := cylinderplot(sin(t)*z^2, t = 0..2*Pi, z = 0..1, axes = boxed):
  s2 := cylinderplot([r, t, 2*r], r = 0..1, t = 0..2*Pi, axes = boxed):

- s1;

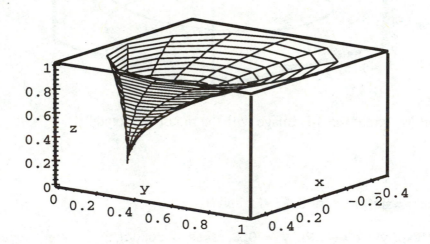

- s2;

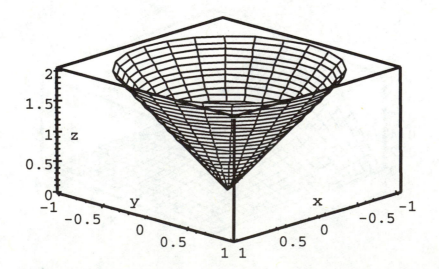

Example 8

If spherical coordinates are defined by the equations $x = r \cos(t) \sin(s)$, $y = r \cos(t) \cos(s)$, $z = r \cos(s)$, then a surface described in spherical coordinates by $r = r(t,s)$ can be graphed by Maple's **sphereplot** command in the plots package. Plot the upper hemisphere by taking $r = 1$, $0 \leq t \leq 2\pi$, $0 \leq s \leq \pi/2$. The option *scaling* must be set to *constrained* to obtain an undistorted hemisphere.

- sphereplot(1, t = 0..2*Pi, s = 0..Pi/2, scaling = constrained, axes = boxed);

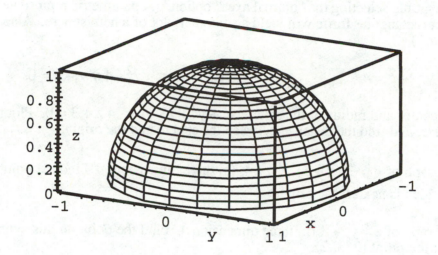

Exercises

Maple supports several possibilities for 3-D plots, including parametric form, rectangular form, and the commands **cylinderplot**, **display3d**, **implicitplot3d**, **pointplot**, **spacecurve**, and **sphereplot**.

1.   Find the center and radius of the sphere. Plot the graph of the upper hemisphere and then the entire graph, selecting the "normal axes" option. (A parametric representation, instead of a direct rectangular form, will yield a smoother plot of a hemisphere. Also, see ?sphereplot.)

   (a) $x^2 + y^2 + z^2 = 4$ $\qquad\qquad$ **(b)** $x^2 + y^2 + z^2 - 4y + 4z = 0$.

2.   Find the center and radius of the sphere $x^2 + y^2 + z^2 - 4x - 4z + 3 = 0$. Plot the lower hemisphere, and find the distance between the sphere and the origin.

3.   Plot the graph of $y = \frac{1}{3}(x + 1)\sqrt{x + 4}$ (in three dimensions). Find the point on this surface which is nearest the origin.

4.   Plot the graph of $z = y^2 + 1$ (in three dimensions). Find the point on this cylinder which is nearest the point $(2, 3, 2)$.

5.   Show that any vector in R^3 can be written as a linear combination of the vectors $[-1, -1, 1]^t$, $[1, 0, 1]^t$ and $[-1, 2, 1]^t$. (These three vectors are mutually orthogonal.)

6.   Can each vector in R^3 be written as a linear combination of the vectors

   (a) $[1, 0, 0]^t$, $[1, 1, 0]^t$ and $[1, 1, 1]^t$?

   (b) $[-1, -1, 1]^t$, $[-1, 1, -1]^t$ and $[1, -1, -1]^t$?

7.   Find the cross product **u** x **v**.(See the linalg package.)

   (a) $\mathbf{u} = [2, 2, 1]^t$, $\mathbf{v} = [1, 4, 1]^t$

   (b) $\mathbf{u} = [3, 1, 2]^t$, $\mathbf{v} = [1, 3, -1]^t$

   (c) $\mathbf{u} = [-1, 0, 0]^t$, $\mathbf{v} = [0, 2, 0]^t$.

8. Find the angle between the vectors $\mathbf{u} = [1, 3, 1]^t$ and $\mathbf{v} = [-1, 1, 2]^t$

   (a) in radians   (b) in degrees.

9. Prove that

   (a) $\mathbf{u} \times \mathbf{v} = -(\mathbf{v} \times \mathbf{u})$ (b) $(\mathbf{u} + \mathbf{v}) \times \mathbf{w} = (\mathbf{u} \times \mathbf{w}) + (\mathbf{v} \times \mathbf{w})$.

10. Find the angle between the vector $\mathbf{u} = [2, 2, 3]^t$ and the plane determined by the vectors $\mathbf{v} = [1, 3, 1]^t$ and $\mathbf{w} = [-2, 1, 2]^t$.

11. Define $\mathbf{v}_1 = [\frac{1}{10}, 2t, -t]^t$ and $\mathbf{v}_2 = [2, t^2, 2]^t$.

   (a) Find all values of t for which $\mathbf{v}_1$ and $\mathbf{v}_2$ are orthogonal.

   (b) Find the angle between $\mathbf{v}_1$ and $\mathbf{v}_2$ when t = 5, 10, 20, 40, 80.

   (c) Let θ be the angle between $\mathbf{v}_1$ and $\mathbf{v}_2$. Find $\lim_{t \to \infty} \theta$.

12. Define $\mathbf{v}_1 = [t^2, -6, \frac{1}{10}]^t$ and $\mathbf{v}_2 = [1, \frac{1}{10}t, -\frac{11}{2}]^t$.

   (a) Find all values of t for which $\mathbf{v}_1$ and $\mathbf{v}_2$ are orthogonal.

   (b) Find the angle between $\mathbf{v}_1$ and $\mathbf{v}_2$ when t = 5, 10, 20, 40, 80.

   (c) Let θ be the angle between $\mathbf{v}_1$ and $\mathbf{v}_2$. Find $\lim_{t \to \infty} \theta$.

13. Plot a segment of the following line in three-space.

   (a) $\mathbf{r}(t) = [2, -1, 3]^t + t[2, 4, -3]^t$ (b) $\mathbf{r}(t) = [3, 1, -1]^t + t[-1, 1, 2]^t$.

14. (a) In the same window, plot the lines $\mathbf{r}_1(t) = [0, 1, 1]^t + t[1, 2, 1]^t$ and
       $\mathbf{r}_2(t) = [5, 1, 6]^t + t[2, -1, 2]^t$.

   (b) Find the intersection point of $\mathbf{r}_1(t)$ and $\mathbf{r}_2(t)$.

15. (a) In the same window, plot the point P = (3, 0, 5) and the line
       $\mathbf{r}(t) = [0, 1, 1]^t + t[1, 2, 1]^t$. If the point P is hard to see, can you enhance the plot?

(b) Find the minimum distance from P to **r**(t).

16. **(a)** Display the planes x - y + 2 z = 1 and 2 x + 3 z = 0.

    (b) Find the angle between the planes. (Use their normal vectors.)

17. **(a)** Display the point P = (2, 5, 4) and the plane x - y + 2 z = 1 in the same window.

    (b) Find the distance between P and each of the following points in the given plane:

b_1) (1, 5,) b_4) (2, 5,)
b_2) (1, 6,) b_5) (2, 6,)
b_3) (1, 7,) b_6) (2, 7,).

    (c) Which point in part **(b)** is closest to P? In the given plane, can you find a point closer to P? Can you find the point in the given plane which is the closest to P?

18. **(a)** Display the point P = (3, 3, 9) and the plane 2 x + y - z = 2 in the same window.

    (b) Find the distance between P and each of the following points in the given plane:

b_1) (3, 3,) b_4) (4, 4,)
b_2) (3, 4,) b_5) (5, 3,)
b_3) (4, 3,) b_6) (5, 4,).

    (c) Which point in part **(b)** is closest to P? In the given plane, can you find a point closer to P? Can you find the point in the given plane which is the closest to P?

19. **(a)** Find the distance between the point P on the line $r_1(s) = [1, 0, 2]^t + s[2, 1, 1]^t$ and the point Q on the line $r_2(t) = [0, 1, 1]^t + t[1, 2, 1]^t$:

    (b) Define the function D(s, t) as the distance between P(s) on $r_1(s)$ and Q(t) on $r_2(t)$. Plot D(s, t) and look for the minimum value.

20. **(a)** Find the distance between the point P on the line $r_1(s) = [2, -3, 1]^t + s[1, 2, 3]^t$ and the point Q on the line $r_2(t) = [1, 1, 1]^t + t[3, -2, 1]^t$:

a_1) P(s = 0.5), Q(t = 0.8) a_3) P(s = 0.7), Q(t = 0.8)
a_2) P(s = 0.5), Q(t = 1) a_4) P(s = 0.7), Q(t = 1).

    (b) Define the function D(s, t) as the distance between P(s) on $r_1(s)$ and Q(t) on $r_2(t)$. Plot D(s, t) and look for the minimum value.

21. Make a movie showing a point moving upward along a circular helixical path.

22. Plot the following three-dimensional surfaces: (In some of these problems a parametric form will produce a somewhat different appearance.)

   (a) $x^2 + y^2 = 1$ **(e)** $4x^2 + z^2 - 8x - 4y - 4z + 16 = 0$

   (b) $y^2 + z^2 = 4$ **(f)** $z = \sqrt{x^2 + y^2 - 4}$

   (c) $x^2 + y^2 + z^2 = 9$ **(g)** $z = \sqrt{x^2 + y^2 + 4}$.

   (d) $z = \sqrt{4 - 4x^2 - \frac{1}{2}y^2}$

23. Plot the following three-dimensional surfaces:

   (a) $r = 2 + 2\cos\theta$ **(f)** $\rho = 4\sin\theta$

   (b) $r = 2 + 3\cos\theta$ **(g)** $\rho = 4\sec\theta$

   (c) $r = 1$ **(h)** $\rho = 1 - \cos\phi$

   (d) $\rho = 1$ **(i)** $\rho = 1 - \sin\phi$

   (e) $\rho = 4\cos\theta$ **(j)** $\rho = 1 - \sin 2\phi$.

Projects

1. Find a relationship between a, b, c, and d which implies that
   $x^2 + y^2 + z^2 + ax + by + cz + d = 0$ represents a sphere of radius > 0.

2. S_1 is the sphere with center at (-4, 4, 5) and radius = 3. S_2 is the sphere with center at (4, 2, -2) and radius = 1.

   (a) In the same window, plot the lower hemisphere of S_1 and the upper hemisphere of S_2. (Get smooth plots - no jagged edges.)

   (b) Find the distance between S_1 and S_2.

3. Define $\mathbf{v}(t) = [t, \sin 2\pi t]^t$.

(a) In the same window, display $\{v(t) \mid t = \frac{k}{8}, k = 1..7\}$.

(b) Define l(t) to be the length of **v**(t). Plot l(t) for $0 \le t \le 1.5$.

(c) Find the vector of maximum length for $0 \le t \le 1$; for $0 \le t \le 4$.

(d) Generate an animation that displays a number of frames of **v**(t) for $0 \le t \le 1$.

4.   Repeat Problem 3 for **v**(t) = $[\sin \pi t, \cos 2\pi t, \frac{3t}{t-2}]^t$.

5.   Generate a reasonable version of the object shown or specified. Give your equations and show your work if a point is specified.

(a) The Mushroom   **(b)** The Top with Point = $(4, \frac{\pi}{2}, \frac{\pi}{3})$

(c) The Toadstool with Point = (0, 3, 2.2)   **(d)** The Star Trek Track

   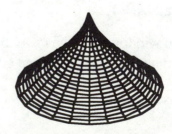

(e) The Vented Sombrero   **(f)** The Mortar Board Cap

(g) The Mortar Board Cap with the board tilted 30° to the horizontal

(h) The Spike, $0 \leq \rho \leq 1$, $1.2 \leq \theta \leq \frac{\pi}{2}$, $1.2 \leq \phi \leq \frac{\pi}{2}$

(i) The Severed Spike, $1 \leq \rho \leq 2$, $1.2 \leq \theta \leq \frac{\pi}{2}$, $1.2 \leq \phi \leq \frac{\pi}{2}$.

6.   Define S_1 to be the sphere $\rho = 7$. Let v_1 be a vector from the origin through the point
     (2, 6, 3). Define v_2 to be the vector parallel to v_1, having initial point at (2, 6, 3) and
     length 4. In the same window, display S_1 and v_2. Is v_2 "perpendicular to" S_1?

7.   With my eye in the center of a cubical room, each wall will occupy one-sixth of my total
     viewing area. If I move halfway toward the center of a wall, what fraction of my total
     viewing area will <u>that wall</u> then occupy? (A Monte Carlo approach is reasonable.)
                                                             -Byron R. Adams <u>(The Bent)</u>

8.   Put a point representing Uppsala at latitude 60° N on a sphere representing Earth, and
     make a movie showing the sphere as it rotates about its' vertical axis. Make the rate of
     rotation approximately 360° every 10 seconds.

9.   Modify your work in Problem 8 so that the movie includes a satellite with an equatorial
     orbit. Make the altitude of the satellite six times the radius of the sphere and set the speed
     of the satellite so that it circles the sphere twice during the time the sphere rotates
     through 360° . (As an extra challenge, see if you can tilt the satellite orbit through 30°
     about a diameter through the Prime Meridian. Uppsala is at longitude 18° E.)

Chapter 15

Vector Functions

New Maple Commands for Chapter 15

with(linalg);	loads linear algebra package
evalm(v/c);	performs matrix arithmetic, here, dividing vector **v** by scalar c
norm(v,2);	computes euclidean norm (length) of vector v

Introduction

The term *vector function* must be interpreted carefully since it can mean a scalar-valued function of a vector argument, a vector-valued function of a scalar argument, or a vector valued function of a vector argument. An example of a scalar-valued function of a vector argument would be the length of the vector. To each vector is associated a scalar, the vector's length.

On the other hand, the term "vector-valued function" refers to a function which produces a vector. Its argument can be either a scalar or a vector. For example, when a space curve, defined parametrically via equations of the form $x = x(t)$, $y = y(t)$, $z = z(t)$, is represented in vector notation, we have the prototypical vector-valued function of the scalar argument, the parameter on the curve. In this representation, the three parametric functions are taken as the components of the radius vector $\mathbf{R} = x(t)\,\mathbf{i} + y(t)\,\mathbf{j} + z(t)\,\mathbf{k}$, so that the space curve is traced by the tip of $\mathbf{R}$ as t varies. In general, such a "vector-valued function" of the scalar t will appear, in this chapter, as $\mathbf{F} = [x(t), y(t), z(t)]^t$, with the superscript indicating "transpose."

Finally, the reader is alerted to the vector-valued function of a vector argument, sometimes written as $\mathbf{F}(\mathbf{X})$, where both $\mathbf{F}$ and $\mathbf{X}$ represent vectors. Physically, such a quantity could be the velocity vector at every point in a three dimensional space. In component form, $\mathbf{F}(\mathbf{X})$ would be the vector $[f(x,y,z), g(x,y,z), h(x,y,z)]^t$. We leave for a course in Advanced Calculus the ramifications of the vector-valued function of the vector argument, concentrating here on the scalar functions of a vector argument, and vector-valued functions of a scalar argument.

Examples

Example 1

Determine if the space curves defined by the vectors $\mathbf{F} = [1+t,\, 2 - 3t,\, 3 + 2\,t]^t$ and $\mathbf{G} = [3 - s/2,\, 2 - 3s/2,\, 1 + 2s]^t$ intersect, and compute their point of intersection if they do.

It is crucial that different parameters be used if the question is merely "Do the curves intersect." If the question were, "Do the curves intersect for the exact same value of the parameter on each curve," then we would use the same parameter in each curve. This distinction is equivalent to

241

interpreting each space curve as the con-trail of a flying jet and asking if observing two intersecting con-trails means the passing jets had collided. Thus, the jets could have passed through the same point in space but at different times so the con-trails intersect, but the jets didn't crash.

Making no apology that we have used two lines in space for this initial example, we set about deciding whether or not these lines intersect.

- with(linalg):
- F := vector([1 + t, 2 - 3*t, 3 + 2*t]);
  G := vector([3 - s/2, 2 - 3*s/2, 1 + 2*s]);

$$F := [1 + t \quad 2 - 3\,t \quad 3 + 2\,t]$$

$$G := \left[3 - \frac{1}{2}s \quad 2 - \frac{3}{2}s \quad 1 + 2\,s \right]$$

We set the first and second components of **F** and **G** equal, and solve for values of t and s that make x and y have equal values in **F** and **G**. We then test to see what the respective z-values at these parameter values are and decide whether or not the space curves intersect.

- q := solve({F[1] = G[1], F[2] = G[2]}, {t, s});

$$q := \{ t = 1, s = 2 \}$$

- subs(q, F[3]);
  subs(q, G[3]);

$$5$$

$$5$$

Since the z-values agree when the x- and y-values agree, these two space curves intersect.

It is instructive to note that the curves do not intersect at the same time and that failure to use two different parameters would have led to the wrong conclusion about intersections.

- solve(F[1] = subs(s=t, G[1]), t);

$$\frac{4}{3}$$

- subs(t=4/3, F[2]);
  subs(s=4/3, G[2]);

$$-2$$

0

Thus, naively using the same parameter on both curves could lead to an erroneous conclusion.

Example 2

Find a vector-valued function whose graph traverses the right branch of the hyperbola $x^2/9 - y^2 = 1$ from the point $(3, 0)$ to the point $(6, \sqrt{3})$.

This question requires that we find a parametric representation of the given parabola. One way to determine a parametrization of a plane curve is to declare that $x = x(t)$ for some function $x(t)$ and to use the cartesian representation of the curve to solve for the companion function $y = y(t)$. However, it is sometimes wiser to use properties of the cartesian formula to obtain particularly effective parametrizations. In this case, we recall that $\sec^2 t - \tan^2 t = 1$ and are inspired to write $\mathbf{R} = [3 \sec(t), \tan(t)]^t$ as the parametrization.

- plot([3*sec(t), tan(t), t = 0..Pi/3]);

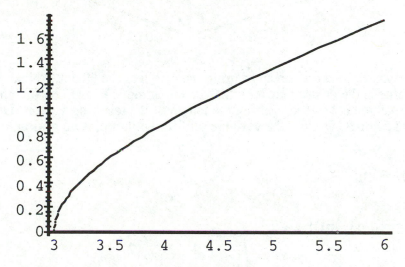

The reader is encouraged to find alternative parametrizations, perhaps by recalling another identity with the difference of squares equaling 1.

Example 3

Obtain the arc length of the space curve given by $\mathbf{R} = [t, t^2, \sin(t)]^t, 0 \le t \le 1$.

First, we sketch the space curve with Maple's **spacecurve** command found in the plots package. We'll enter the vector-valued function $\mathbf{R}$ as a Maple vector and use the **op** function to extract the list of functions where needed. (Note that a Maple vector is formed by applying the command **vector** to a list of components. The **op** command extracts the list from the vector.)

- with(plots):
- R := vector([t, t^2, sin(t)]);

$$R := \begin{bmatrix} t & t^2 & \sin(t) \end{bmatrix}$$

- spacecurve(op(R), t = 0..1, axes = boxed, labels = [x, y, z]);

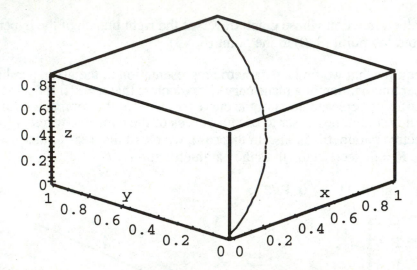

The graph above was rotated on-screen to bring the origin to the "front." Next, we set up an integral whose value is the desired arc length. The integrand is formed by differentiating each component of the vector **R** and then getting the length of the resulting vector. Differentiation is mapped onto the vector **R**, while t, the variable of differentiation, is an option to the **map** command.

- q1 := map(diff, R, t);

$$q1 := \begin{bmatrix} 1 & 2t & \cos(t) \end{bmatrix}$$

- q2 := sqrt(dotprod(q1, q1));

$$q2 := \sqrt{1 + 4t^2 + \cos(t)^2}$$

- q3 := Int(q2, t = 0..1);

$$q3 := \int_0^1 \sqrt{1 + 4t^2 + \cos(t)^2} \, dt$$

- evalf(q3);

$$1.728655691$$

Example 4

Find a unit tangent vector along the curve defined by **R** in Example 3.

A fundamental idea in vector calculus is that differentiating a vector-valued function with respect to its parameter yields a tangent vector. The reader need only think of a spacecurve as defining the path of a moving particle and note that the velocity vector is along the motion. Hence, we have already seen how to obtain a tangent vector to **R**. It is the vector q1 found in Example 3. All that is left is the normalization, which could be done by the **normalize** command in the linear algebra package.

• normalize(q1);

$$\left[\frac{1}{\sqrt{1 + 4|t|^2 + |\cos(t)|^2}} \quad 2\frac{t}{\sqrt{1 + 4|t|^2 + |\cos(t)|^2}} \quad \frac{\cos(t)}{\sqrt{1 + 4|t|^2 + |\cos(t)|^2}} \right]$$

The vector q1 has been divided by its 2-norm, which is the "length" of q1 found by taking the square root of the sum of the squares of the components. Unfortunately, the 2-norm leaves us with absolute values that could make succeeding computations more difficult. In fact, note

• norm(q1, 2);

$$\sqrt{1 + 4|t|^2 + |\cos(t)|^2}$$

That is why we used the dot product in Example 3 when we obtained q2 for the length of the vector q1. Even though we could have handled the absolute value difficulty by assuming t positive, we opt to constuct the unit tangent vector **T** via

• T := evalm(q1/q2);

$$T := \left[\frac{1}{\sqrt{1 + 4t^2 + \cos(t)^2}} \quad 2\frac{t}{\sqrt{1 + 4t^2 + \cos(t)^2}} \quad \frac{\cos(t)}{\sqrt{1 + 4t^2 + \cos(t)^2}} \right]$$

Example 5

Compute the curvature along the space curve given by **R** in Example 3.

A curvature formula compatible with a curve given parametrically is k = ||**R**' x **R**"||/||**R**'||³.

We already have **R** entered as R, **R**' as q1, and ||**R**'|| as q2. Thus,

• R2 := map(diff, q1, t);

$$R2 := [0 \quad 2 \quad -\sin(t)]$$

- q4 := crossprod(q1, R2);

$$q4 := [-2\,t\sin(t) - 2\cos(t) \quad \sin(t) \quad 2]$$

- q5 := sqrt(dotprod(q4, q4));

$$q5 := \sqrt{(-2\,t\sin(t) - 2\cos(t))^2 + \sin(t)^2 + 4}$$

- q6 := q5/q2^3;

$$q6 := \frac{\sqrt{(-2\,t\sin(t) - 2\cos(t))^2 + \sin(t)^2 + 4}}{\left(1 + 4\,t^2 + \cos(t)^2\right)^{3/2}}$$

The expression q6 is the curvature at any point on the curve defined by **R**. For example, at $t = \pi/4$ the curvature is

- q7 := simplify(subs(t = Pi/4, q6));

$$q7 := 2\,\frac{\sqrt{2\,\pi^2 + 16\,\pi + 104}}{\left(6 + \pi^2\right)^{3/2}}$$

- evalf(q7);

$$.4173124324$$

In fact, a graph of q6 shows the curvature all along the curve **R**.

- plot(q6, t = 0..1);

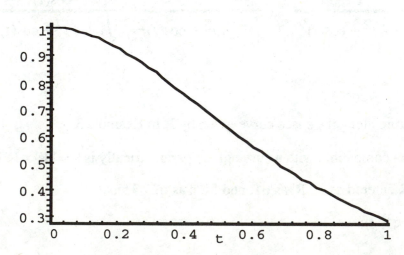

Example 6

Find, for the plane curve $y = x^3$, the circle of curvature that touches the curve at the point (1, 1).

A direct attack on this problem begins with the recognition that the center of curvature will sit at a distance of 1/k from (1, 1) on a line normal to y(x). The strategy this engenders is to obtain the equation of the normal line, obtain the radius of curvature at x = 1, and locate the center of curvature at the appropriate distance along the normal line. First, obtain the radius of curvature as the reciprocal of the curvature.

- q8 := 1/subs(x = 1, diff(x^3,x,x)/(1+diff(x^3,x)^2)^(3/2));

$$q8 := \frac{5}{3}\sqrt{10}$$

Next, find the equation of the normal line through (1, 1). Since y'(1) = 3, the slope of the required normal line is -1/3. Hence,

- q9 := y = -1/3*(x - 1) + 1;

$$q9 := y = -\frac{1}{3}x + \frac{4}{3}$$

We next look along this line for a point (xc, yc) whose distance from (1, 1) is given by q8.

- q10 := sqrt((1 - xc)^2 + (1 - subs(x = xc, rhs(q9)))^2) = q8;

$$q10 := \frac{1}{3}\sqrt{10}\sqrt{(-1 + xc)^2} = \frac{5}{3}\sqrt{10}$$

- q11 := solve(q10, xc);

$$q11 := 6, -4$$

Since, at x = 1, the curve $y = x^3$ is concave up and has a positive slope, the circle of curvature must lie to the left of (1, 1). Hence, we want xc = -4. That means yc is

- yc := subs(x = -4, rhs(q9));

$$yc := \frac{8}{3}$$

The circle with center (xc, yc) = (-4, 8/3) and radius q8 is given parametrically by

- u := -4 + q8*cos(t);
  v := 8/3 + q8*sin(t);

$$u := -4 + \frac{5}{3}\sqrt{10}\,\cos(t)$$

$$v := \frac{8}{3} + \frac{5}{3}\sqrt{10}\,\sin(t)$$

A graph of y(x) and the circle of curvature is given by

• plot({[u, v, t = 0..2*Pi], [t, t^3, t = 0..2], [[1,1], [-4,8/3]]}, scaling = constrained);

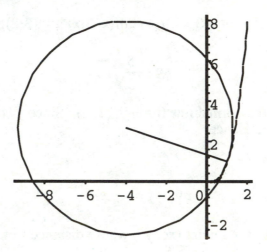

Example 7

Find **N**, the Principal Normal along the curve given by **R** in Example 3.

The Principal Normal is given by the relationship **N** = d**T**/ds / ‖d**T**/ds‖. The variable of differentiation s is arc length; the parameter along **R** is t. To carry out the required differentiation will take the chain rule. Thus, d**T**/ds = (d**T**/dt) (dt/ds) = (d**T**/dt) / (ds/dt). We found ds/dt in Example 3 when we computed the element of arc length to be q2; thus, ds/dt = ‖d**R**/dt‖ = q2. We therefore obtain d**T**/dt and call it Tt in Maple.

• Tt := map(diff, T, t);

$$Tt := \left[-\frac{1}{2}\frac{8\,t - 2\cos(t)\sin(t)}{\left(1 + 4\,t^2 + \cos(t)^2\right)^{3/2}} \quad -\frac{t\,(8\,t - 2\cos(t)\sin(t))}{\left(1 + 4\,t^2 + \cos(t)^2\right)^{3/2}} + 2\,\frac{1}{\sqrt{1 + 4\,t^2 + \cos(t)^2}} \right.$$

$$\left. -\frac{1}{2}\frac{\cos(t)\,(8\,t - 2\cos(t)\sin(t))}{\left(1 + 4\,t^2 + \cos(t)^2\right)^{3/2}} - \frac{\sin(t)}{\sqrt{1 + 4\,t^2 + \cos(t)^2}} \right]$$

Dividing dT/dt by ds/dt yields dT/ds, which we label Ts in Maple.

- Ts := evalm(Tt/q2);

$$Ts := \left[-\frac{1}{2} \frac{8\,t - 2\cos(t)\sin(t)}{\left(1 + 4\,t^2 + \cos(t)^2\right)^2} \quad \frac{-\dfrac{t\,(8\,t - 2\cos(t)\sin(t))}{\left(1 + 4\,t^2 + \cos(t)^2\right)^{3/2}} + 2\,\dfrac{1}{\sqrt{1 + 4\,t^2 + \cos(t)^2}}}{\sqrt{1 + 4\,t^2 + \cos(t)^2}} \right.$$

$$\left. \frac{-\dfrac{1}{2}\dfrac{\cos(t)\,(8\,t - 2\cos(t)\sin(t))}{\left(1 + 4\,t^2 + \cos(t)^2\right)^{3/2}} - \dfrac{\sin(t)}{\sqrt{1 + 4\,t^2 + \cos(t)^2}}}{\sqrt{1 + 4\,t^2 + \cos(t)^2}} \right]$$

We are well advised to simplify dT/ds before attempting to normalize it.

- Ts1 := map(simplify,Ts);

$$Ts1 := \left[-\frac{4\,t - \cos(t)\sin(t)}{\%1} \quad 2\,\frac{t\cos(t)\sin(t) + 1 + \cos(t)^2}{\%1} \right.$$

$$\left. -\frac{4\cos(t)\,t + \sin(t) + 4\sin(t)\,t^2}{\%1} \right]$$

$$\%1 := 1 + 8\,t^2 + 2\cos(t)^2 + 16\,t^4 + 8\,t^2\cos(t)^2 + \cos(t)^4$$

Now we want the magnitude of dT/ds, computed from the form Ts1 in Maple.

- q12 := factor(dotprod(Ts1, Ts1));

$$q12 := \frac{4\sin(t)^2\,t^2 + 8\,t\cos(t)\sin(t) + 4 + 4\cos(t)^2 + \sin(t)^2}{\left(1 + 4\,t^2 + \cos(t)^2\right)^3}$$

We anticipated the complexity of the magnitude and factored before taking the square root.

• q13 := sqrt(q12);

$$q13 := \sqrt{\frac{4\sin(t)^2\, t^2 + 8\, t\cos(t)\sin(t) + 4 + 4\cos(t)^2 + \sin(t)^2}{\left(1 + 4\, t^2 + \cos(t)^2\right)^3}}$$

The expression in q13 is the length of the vector d**T**/ds from which we obtain **N**, the Principal Normal vector, by dividing d**T**/ds by q13, its length.

• N := evalm(Ts1/q13);

$$N := \left[-\frac{4\, t - \cos(t)\sin(t)}{\sqrt{\%2}\ \%1}\quad 2\,\frac{t\cos(t)\sin(t) + 1 + \cos(t)^2}{\sqrt{\%2}\ \%1} \right.$$

$$\left. -\frac{4\cos(t)\, t + \sin(t) + 4\sin(t)\, t^2}{\sqrt{\%2}\ \%1} \right]$$

$$\%1 := 1 + 8\, t^2 + 2\cos(t)^2 + 16\, t^4 + 8\, t^2\cos(t)^2 + \cos(t)^4$$

$$\%2 := \frac{4\sin(t)^2\, t^2 + 8\, t\cos(t)\sin(t) + 4 + 4\cos(t)^2 + \sin(t)^2}{\left(1 + 4\, t^2 + \cos(t)^2\right)^3}$$

We make one attempt to tidy up the expression for **N**. We do this by both simplifying and factoring. Several operations can be mapped onto a vector if these operations are seen as the names of Maple functions. Since sequencing operators one after the other is actually composition, we are able to factor after simplifying by the composition (via the @ symbol) in the following command.

• map(factor @ simplify, N);

$$\left[-\frac{4\, t - \cos(t)\sin(t)}{\sqrt{\%1}\,\left(1 + 4\, t^2 + \cos(t)^2\right)^2}\quad 2\,\frac{t\cos(t)\sin(t) + 1 + \cos(t)^2}{\sqrt{\%1}\,\left(1 + 4\, t^2 + \cos(t)^2\right)^2} \right.$$

$$\left. -\frac{4\cos(t)\, t + \sin(t) + 4\sin(t)\, t^2}{\sqrt{\%1}\,\left(1 + 4\, t^2 + \cos(t)^2\right)^2} \right]$$

$$\%1 := -\frac{-8\, t\cos(t)\sin(t) - 5 - 3\cos(t)^2 + 4\, t^2\cos(t)^2 - 4\, t^2}{\left(1 + 4\, t^2 + \cos(t)^2\right)^3}$$

We conclude by observing, finally, that the length of d**T**/ds is actually the curvature k. Since we computed the curvature as q6 in Example 5, we test k^2 against q12, the square of the length of d**T**/ds.

• simplify(q6^2 - q12);

$$0$$

Since the difference is zero, we conclude that we have verified the relationship k = ‖d**T**/ds‖. Hence, **N** = (d**T**/ds)/k.

Exercises

1.   Plot each vector-valued function, showing two views of each plot.

(a) $r(t) = [\cos t, \sin t, t]^t, 0 \le t \le 3\pi$ (b) $r(t) = [\cos t, 2 \sin t, t]^t, 0 \le t \le 2\pi$.

2.   Define $r_1(t) = [1 + t, 2 - 3t, 3 + 2t]^t$ and $r_2(s) = [3-s/2, 2 - 3s, 1 + 2s]^t$. Make a movie showing the two vector-valued functions as $r_1(t)$ moves from $r_1(0)$ to $r_1(2)$ and $r_2(s)$ moves from $r_2(0)$ to $r_2(2)$. You should see that the two spacecurves share a common point, not reached at the same time.

3.   Define
$$r_1(t) = [\cos 2t, 2 \sin t, 1 + \frac{t^3}{10}]^t \text{ and } r_2(t) = [-1 + 0.971t, 1 + 0.435t, 1 + 0.051t]^t.$$

(a) In the same window plot $r_1(t)$ and $r_2(t)$. Do the curves appear to intersect? Look at the plot from several different viewpoints.

(b) Find all intersection points of the two spacecurves.

4.   Make a movie of the two curves defined in Exercise 3 as the parameter value of each curve increases from 0 to 2.

5.   Define
$$r_1(t) = [\cos 2t, 2 \sin t, 1 + \frac{t^3}{10}]^t \text{ and } r_2(t) = [-2.227 + 0.4t^3, 0.344t^2, 0.470 + 0.3t]^t.$$

(a) In the same window plot $r_1(t)$ and $r_2(t)$. Do the curves appear to intersect? Look at the plot from several different viewpoints.

(b) Find all intersection points of the two spacecurves.

6.   Define $r(t) = [\cos t + t \sin t, \sin t - t \cos t]^t$.

(a) Find a tangent vector to the graph of $r(t)$ at the point where $t = \frac{\pi}{4}$.

(b) In the same window plot $r(t)$ and your tangent vector.

7.   Define $r(t) = [t^3 - 3t, t^2, t]^t$.

(a) Find unit vectors tangent to $r(t)$ at the points where $t = 1$, and then $t = 2$.

(b) Find unit vectors normal to **r**(t) at the points where t = 1, and then t = 2.

(c) In the same window, plot **r**(t) and the four vectors from parts **(a)** and **(b)**.

8. Find a vector-valued function whose graph satisfies the following conditions, and then plot the graph.

(a) Traverses a circle of radius 1 in a counterclockwise direction, starting at the rightmost point.

(b) Traverses a circle of radius 4 in a clockwise direction, starting at the rightmost point.

(c) Traverses an ellipse, major axis horizontal and of length 10, minor axis of length 4, starting at the rightmost point and moving counterclockwise.

(d) Traverses an ellipse, major axis vertical and of length 8, minor axis of length 6, starting at the lowest point and moving clockwise.

9. Find a vector-valued function whose graph has the shape shown. Plot the graph of your function.

(a)   **(b)**

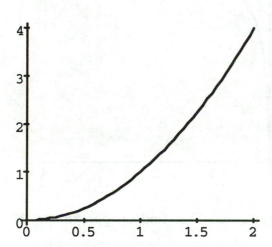   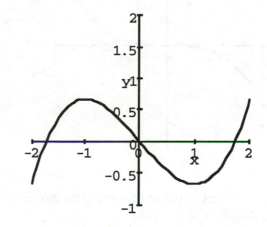

(c)

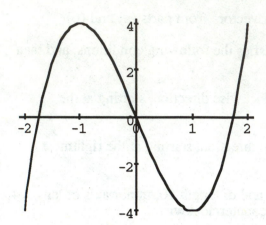

(d)

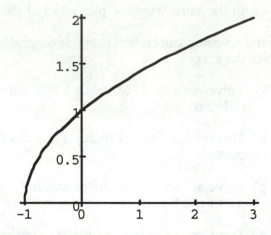

(e)

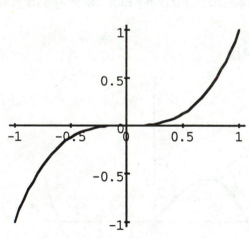

(f)

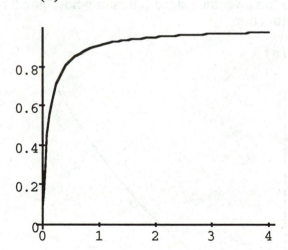

10. Plot the vector-valued function on the interval specified. Then find, or approximate, the arc length of the curve.

   (a) $\mathbf{r}(t) = [4 \cos t,\ 4 \sin t,\ 3t]^t,\ \ 0 \le t \le \pi$

   (b) $\mathbf{r}(t) = [\cos^3 t,\ \sin^3 t,\ 1]^t,\ \ \ 0 \le t \le \pi$

   (c) $\mathbf{r}(t) = [t \cos t,\ t \sin t,\ t^{3/2}]^t,\ \ 0 \le t \le \pi$

(d) $\mathbf{r}(t) = [\cos^2 t, \sin^2 t, t]^t,$ $\qquad 0 \le t \le \pi.$

11. Plot the vector-valued function on the interval specified. Then find the curvature at each point specified.

(a) $\mathbf{r}(t) = [\cos t + t \sin t, \sin t - t \cos t]^t,$ $\quad [0, \pi/2], \quad t_0 = \dfrac{\pi}{4}$

(b) $\mathbf{r}(t) = [t, \ln(\sec t)]^t,$ $\quad [0, 1.4], \quad t_0 = \dfrac{\pi}{3}, \; t_0 = 1.4$

(c) $\mathbf{r}(t) = [\cos t, \sin t, t]^t,$ $\quad [0, 2\pi], \quad t_0 = \dfrac{\pi}{2}, \; t_0 = \dfrac{3\pi}{4}$

(d) $\mathbf{r}(t) = [1 + t, 2 + 2t, 3 + 3t]^t,$ $\quad [-1, 2], \quad t_0 = 0, \; t_0 = 1$

(e) $\mathbf{r}(t) = [e^t \cos t, e^t \sin t, t^2]^t,$ $\quad [0, 3], \quad t_0 = 1, \; t_0 = 2$

(f) $\mathbf{r}(t) = [\cosh t, \sinh t, t]^t,$ $\quad [-1..3], \quad t_0 = 0, \; t_0 = 2.$

12. In the same window, plot a section of the curve near the point where $t = t_0$, and show the osculating circle at that point.

(a) $\mathbf{r}(t) = [t, t^2]^t,$ $\qquad t_0 = 1$

(b) $\mathbf{r}(t) = [t, \cos t]^t,$ $\qquad t_0 = \dfrac{\pi}{3}$

(c) $\mathbf{r}(t) = [t, \ln(\sec t)]^t,$ $\qquad t_0 = \dfrac{\pi}{3}$

(d) $\mathbf{r}(t) = [t, 60t - 48t^2 + 12t^3 - t^4]^t,$ $\qquad t_0 = 1$

(e) $\mathbf{r}(t) = [\cos t + t \sin t, \sin t - t \cos t]^t,$ $\qquad t_0 = \dfrac{\pi}{4}.$

13. Find the minimum radius of curvature of the curve, and plot a section of the curve and its osculating circle at that point.

(a) $\mathbf{r}(t) = [2t, t^2 - 1]^t$

(b) $\mathbf{r}(t) = [t, 3t^r - 4t^3 - 24t^2 + 50t + 2]^t.$

14. A particle moves according to $\mathbf{r}(t) = [t - t^3, t^2, \sin t]^t$. Find the minimum speed of the particle for $0.1 \leq t \leq 1$.

15. A particle moves according to $\mathbf{r}(t) = [\ln t, \cos t, \frac{1}{t}]^t$.

   (a) Find the minimum speed for $2 \leq t \leq 5$; $2 \leq t \leq 8$; $2 \leq t \leq 10$.

   (b) Find the maximum speed for $4 \leq t \leq 10$.

Projects

1. Define $\mathbf{r}(t) = [t, \sin t]^t$ and let $t_0 = 1.4$.

   (a) Let $\mathbf{v}_k$ be the vector having initial point at $\mathbf{r}(t_0)$ and terminal point at $\mathbf{r}(t_0 + \Delta t_k)$. Take $\Delta t_1 = 0.35$, $\Delta t_2 = 0.20$, and $\Delta t_3 = 0.05$. Note that these values define $\mathbf{v}_1$, $\mathbf{v}_2$, and $\mathbf{v}_3$. Plot, in the same window, $\mathbf{r}(t)$ for $1 \leq t \leq 2$, $\mathbf{v}_1$, $\mathbf{v}_2$, and $\mathbf{v}_3$.

   (b) Let $\mathbf{u}_k$ be the unit vector having initial point at $\mathbf{r}(t_0)$ and direction toward $\mathbf{r}(t_0 + \Delta t_k)$. Take $\Delta t_1 = 0.35$, $\Delta t_2 = 0.20$, and $\Delta t_3 = 0.05$. Plot, in the same window, $\mathbf{r}(t)$ for $1 \leq t \leq 2$, $\mathbf{u}_1$, $\mathbf{u}_2$, and $\mathbf{u}_3$.

   (c) Find a unit tangent vector to $\mathbf{r}(t)$ at $\mathbf{r}(t_0)$, using the idea of part (b) and taking an appropriate limit.

   (d) Use the derivative to find a unit tangent vector to $\mathbf{r}(t)$ at $\mathbf{r}(t_0)$. Compare with your result in part (c).

2. Define $\mathbf{r}(t) = [\cos t, \sin t, \sqrt{t}]^t$. Repeat parts (a) through (d) of Problem 1.

3. Define $\mathbf{r}_1(t) = [\frac{1}{3}t^3, \frac{1}{2}t^2]^t$.

   (a) Find the arc length parameter representation $\mathbf{r}_2(s)$ of $\mathbf{r}_1(t)$. (Measured from $t = 0$.)

   (b) Find s_0 corresponding to $t_0 = 1.5$. Thus, solve $\mathbf{r}_1(t_0) = \mathbf{r}_2(s_0)$. Then compute the tangent vectors $\mathbf{v}_1 = \mathbf{r}_2'(s_0)$ and $\mathbf{v}_2 = \mathbf{r}_2'(s_0 + 0.1)$.

   (c) Let kap = one tenth of the angle between $\mathbf{v}_1$ and $\mathbf{v}_2$. Show that kap is a (crude) measure of how fast the curve is changing direction.

(**d**) Find the curvature κ of $\mathbf{r}_1(t)$ at the point corresponding to $t_0 = 1.5$. Compare your answer with the value of kap from part (**c**).

4. (**a**) Write Maple instructions for computing the curvature of a vector-valued function at a given point. Telling Maple the t value corresponding to the point at which the curvature is to be found should then allow the appropriate value of the curvature to be found.

(**b**) Use your Maple code to find the curvature, at $t = \frac{\pi}{6}$ and $t = 1$ for

$\mathbf{r}(t) = [\cos t , \sin t , t]^t$.

(**c**) Use your Maple code to find the curvature at $t = \frac{\pi}{3}$ and $t = 0$ for

$\mathbf{r}(t) = [\cos t , \sin t , \sqrt{t}\,]^t$.

5. Plot the graph of the curvature function $\kappa(t)$ of $\mathbf{r}(t)$ on the given interval. Then find the maximum and minimum curvature of $\mathbf{r}(t)$ on the given interval.

(**a**) $\mathbf{r}(t) = [4 \cos t, \sin t]^t$, $0 \le t \le \pi$

(**b**) $\mathbf{r}(t) = [4 \cos t, 4 \sin t, 2t]^t$, $0 \le t \le 3\pi$

(**c**) $\mathbf{r}(t) = [4 \cos t, 4 \sin t, t^2]^t$, $0 \le t \le 3\pi$

(**d**) $\mathbf{r}(t) = [\cos t, \sin t, \ln t]^t$, $0.1 \le t \le 100$

(**e**) $\mathbf{r}(t) = [\cos t, 2 \sin t, e^{-t}]^t$, $0 \le t \le 4\pi$

(**f**) $\mathbf{r}(t) = [t, \frac{1}{t}, \sqrt{t}\,]^t$, $0.1 \le t \le 10$.

6. (**a**) For the given vector-valued function $\mathbf{r}(t)$, plot the points corresponding to $t = \frac{k}{20}$, for $k = 1..40$. How do these points indicate the relative speed of the particle whose position is described by $\mathbf{r}(t)$?

(**b**) Print the value of the speed of the particle at $t = \frac{k}{20}$, for $k = 1..40$. Do these numbers agree with the information from the plot in part (**a**)?

i) $\mathbf{r}(t) = [t, \frac{2}{t}]^t$

ii) $\mathbf{r}(t) = [t, \frac{2}{t}, t^{1.5}]^t$

       iii) $\mathbf{r}(t) = [\cos(3\pi t), \sin(\pi t), t^2]^t$.

7.   Two spacecurves are shown, each with two views. Generate a reasonable version of each and show your Maple commands.

    (a)

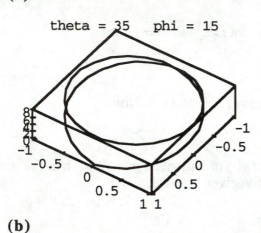 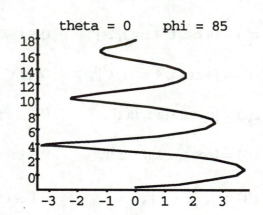

    (b)

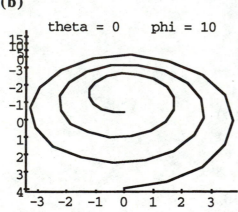

8.   Define $\mathbf{r}_1(t) = [\cos t, \sin t, t]^t$ and $\mathbf{r}_2(t) = [\cos t, \sin t, t^2]^t$.

    (a) Which curve is turning more sharply at $t = 3$?

    (b) On which is the speed greater at $t = 3$?

    (c) Does higher speed "straighten out" the curve?

9.   Define $\mathbf{r}_1(t) = [t, \dfrac{1}{\sqrt{\sin t}}, t]^t$ and $\mathbf{r}_2(t) = [t, \dfrac{1}{\sqrt{\sin t}}, \ln t]^t$.

    Repeat parts **(a)**, **(b)**, and **(c)** from Problem 8.

10. Define $\mathbf{r}_1(t) = [\cos^3 t, \sin^3 t]^t$ and $\mathbf{r}_2(t) = [\frac{1}{3}t^3, \frac{1}{2}t^2]^t$.

   (a) Find the first quadrant point of intersection P_0 of $\mathbf{r}_1(t)$ and $\mathbf{r}_2(t)$.

   (b) Find the arc length parameter representations of $\mathbf{r}_1(t)$ and $\mathbf{r}_2(t)$. (Measure from $t = 0$.)

   (c) Find the first quadrant point of intersection of your functions in part (b).

   (d) Which is turning more quickly at P_0, $\mathbf{r}_1(t)$ or $\mathbf{r}_2(t)$?

11. County Road 800 East curves into Maple Ave. southeast as shown. (CR 800 E runs due south and Maple Ave. runs S 45° E.) The curved connecting road x is to satisfy the following conditions:

   (a) AB = BC = 100 meters.

   (b) The tan to x at A is parallel to CR 800.

   (c) The tan to x at C is parallel to Maple Ave.

   (d) The maximum curvature κ of x is to be minimized. (Why is this a practical request?)

   Can you find a good curve to be used for this road?

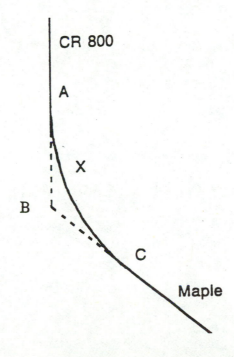

Chapter 16
Partial Derivatives

New Maple Commands for Chapter 16

collect(f,[p,q]);	in f, group all terms with p, and all terms with q
readlib(mtaylor);	access the mtaylor command
mtaylor(f(x,y),[x=a,y=b],n);	compute multivariable Taylor polynomial of degree n, about the point $(x,y) = (a,b)$
with(linalg);	loads linear algebra package
grad(f(x,y,z),[x,y,z]);	returns gradient vector $f_x \mathbf{i} + f_y \mathbf{j} + f_z \mathbf{k}$
with(plots);	loads plots package
contourplot(f(x,y),x=a..b,y=c..d);	returns contour plot of f(x,y)
implicitplot3d(f(x,y,z),x=a..b,y=c..d,z=p..q);	
	plots surface z=z(x,y) defined implicitly by f(x,y,z)

Introduction

The two branches of calculus, Differential Calculus and Integral Calculus, have their counterparts in multivariable calculus. In this chapter we study the meaning and application of differentiation for functions of several variables. Such derivatives are called partial derivatives and have an interpretation similar to that of ordinary derivatives.

Geometrically, when thinking about ordinary derivatives for functions of a single variable, we think of tangents to curves. The analogous image for functions of several variables would be a tangent plane touching a surface. While not needed for defining a partial derivative, the concept of a limit for functions of several variables is distinct enough to warrant a preliminary discussion.

Examples

Example 1

If $f(x,y) = xy/(x^2 + y^2)$, study the limit of f(x,y) as the point (x, y) approaches the origin, (0, 0).

- f := x*y/(x^2 + y^2);

$$f := \frac{x\,y}{x^2 + y^2}$$

Begin by loading the plots package to access the **contourplot** command with which we will generate a contour plot of the surface defined by z = f(x,y). A contour plot is a collection of

261

level curves, that is, curves in the xy-plane on which the function is constant.

- with(plots):

- contourplot(f, x = -1..1, y = -1..1);

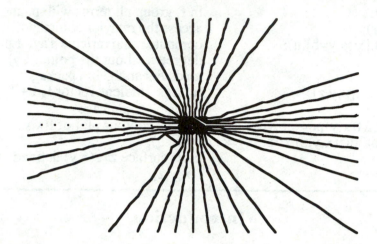

The contour plot suggests that f(x,y) is constant on lines y = mx through the origin. That prompts us to examine f(x, mx).

- fm := subs(y = m*x, f);

$$fm := \frac{x^2\, m}{x^2 + m^2\, x^2}$$

For any x ≠ 0 fm simplifies to

- f1 := simplify(fm);

$$f1 := \frac{m}{1 + m^2}$$

Indeed, fm is constant on lines through the origin. This means that near (0, 0), f(x,y) will have any of the values given by f1, depending on which line y = mx contains the point of evaluation. We therefore say that the limit at (0, 0) depends on the direction of approach taken to the origin. Such a directionally-dependent limit does not exist.

It is far easier to show that a limit does not exist than it is either to find a multivariable limit or to verify one. To show that a limit does not exist, it is enough to show that it is directionally dependent, and that takes only two directions yielding different limits.

Example 2

Obtain a contour plot of the function $f(x,y) = \sin(x)\, e^{-y}$ on the domain $0 \le x \le \pi$, $0 \le y \le 3$.

The following plot shows the surface and the contours in place on the surface. The reader is invited to rotate the surface so that the view is from high up on the z-axis. The resulting view is that of a contour plot.

- f := sin(x)*exp(-y);

$$f := \sin(x)\, e^{(-y)}$$

- plot3d(f, x = 0..Pi, y = 0..3, style = patchcontour, axes = boxed);

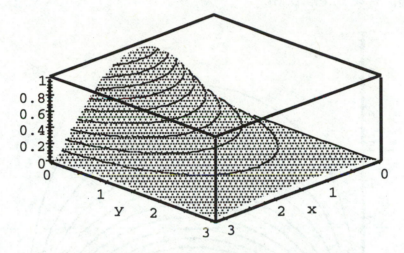

A contour plot can be obtained directly via the **contourplot** command.

- contourplot(f, x = 0..Pi, y = 0..3, axes = normal);

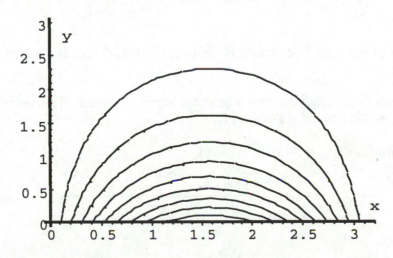

Next, let's follow the definition of a contour plot and create the individual level curves for f. It is clear from f that $z = 0$ along $x = 0$ and $x = \pi$. This helps to avoid problems when we set $f(x,y) = c$, a constant, and solve for $y = y(x)$.

- q := solve(f = c, y);

$$q := -\ln\left(\frac{c}{\sin(x)}\right)$$

Select a spread of values for c and create a separate level curve $y(x)$ for each such value.

- for k from 1 to 10 do y.k := subs(c = k/10, q); od:

- plot({y.(1..10)}, x = 0..Pi, 0..3);

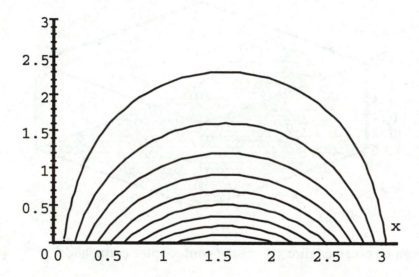

Example 3

Use the definition of the partial derivative $\partial f/\partial x$ to obtain this derivative for the function $f(x,y)$ given in Example 2.

The partial derivative $\partial f/\partial x$ requires that y be considered a constant. This derivative is also neatly referenced by the subscript notation f_x.

- fx := Limit((subs(x=x+h, f) - f)/h, h = 0);

$$fx := \lim_{h \to 0} \frac{\sin(x + h)\, e^{(-y)} - \sin(x)\, e^{(-y)}}{h}$$

• value(fx);

$$\frac{\cos(x)}{e^y}$$

• diff(f, x);

$$e^{(-y)}\cos(x)$$

This partial derivative is really an ordinary derivative in the zx-plane along the curve formed when the surface z = f(x,y) is cut by a plane y = constant. As seen above, the **diff** operator in Maple is a partial differentiation operator since all variables, except the variable of differentiation, are held constant during the differentiation. In fact, if g(x,y) is an unspecified function of the two independent variables x and y, the partial derivative with respect to x is implemented in Maple by

• diff(g(x,y), x);

$$\frac{\partial}{\partial x}g(x,y)$$

Example 4

For the surface of Examples 3 and 4, obtain tangent vectors along the plane sections corresponding to cuts by the planes x = constant and y = constant.

The curve formed on the surface when it is cut by the plane x = c, constant, is given parametrically by x = c, y = y, z = f(c,y). In this case, the parameter along the curve is y. Then a tangent vector is formed by differentiating the curve's functions with respect to the parameter y.

• with(linalg):

• ycurve := vector([c, y, subs(x = c, f)]);

$$ycurve := \begin{bmatrix} c & y & \sin(c)\,e^{(-y)} \end{bmatrix}$$

• ytan := map(diff, ycurve, y);

$$ytan := \begin{bmatrix} 0 & 1 & -\sin(c)\,e^{(-y)} \end{bmatrix}$$

The vector ytan is along ycurve, a curve that lies in the yz-plane x = constant. The corresponding xtan is formed from xcurve, a curve in the xz-plane resulting from a cut of the surface by the plane y = C, constant.

• xcurve := vector([x, C, subs(y = C, f)]);

$$xcurve := \begin{bmatrix} x & C & \sin(x)\,e^{(-C)} \end{bmatrix}$$

• xtan := map(diff, xcurve, x);

$$xtan := \begin{bmatrix} 1 & 0 & \cos(x)\,e^{(-C)} \end{bmatrix}$$

Example 5

Find a vector perpendicular to the surface of Examples 2, 3, and 4.

We already have vectors tangent to two distinct directions that are tangent to the surface. If we construct a new vector that is perpendicular to both of these vectors, this new vector must be normal to the surface itself. The two tangents lie in a plane that is itself tangent to the surface, and the normal vector for the surface is the normal of the tangent plane.

• n := crossprod(xtan, ytan);

$$n := \begin{bmatrix} -\cos(x)\,e^{(-C)} & \sin(c)\,e^{(-y)} & 1 \end{bmatrix}$$

Since c represents a value of x and C represents a value of y, our normal vector would look better if we wrote it as

• N := subs(c = x, C = y, op(n));

$$N := \begin{bmatrix} -\cos(x)\,e^{(-y)} & \sin(x)\,e^{(-y)} & 1 \end{bmatrix}$$

Example 6

Derive a formula for the vector normal to a surface given explicitly by an equation of the form z = f(x,y).

The inspiration for the following derivation is contained in Examples 4 and 5.

• f := 'f':
• xcurve := vector([x, y, f(x,y)]);
  ycurve := vector([x, y, f(x,y)]);

$$xcurve := \begin{bmatrix} x & y & f(x,y) \end{bmatrix}$$

$$ycurve := [x \quad y \quad f(x,y)]$$

- xtan := map(diff, xcurve, x);
  ytan := map(diff, ycurve, y);

$$xtan := \left[1 \quad 0 \quad \frac{\partial}{\partial x} f(x,y) \right]$$

$$ytan := \left[0 \quad 1 \quad \frac{\partial}{\partial y} f(x,y) \right]$$

- N := crossprod(xtan, ytan);

$$N := \left[-\left(\frac{\partial}{\partial x} f(x,y) \right) \quad -\left(\frac{\partial}{\partial y} f(x,y) \right) \quad 1 \right]$$

The result just obtained is that the normal to a surface $z = f(x,y)$ is the vector $[-f_x, -f_y, 1]^t$.

Example 7

At the point (2, 3) obtain the equation of the plane that is tangent to the surface given by $z = x^3 y + 2xy^2 - 5x + 7y$.

- f := x^3*y + 2*x*y^2 - 5*x + 7*y;

$$f := x^3 y + 2 x y^2 - 5 x + 7 y$$

- fx := subs(x = 2, y = 3, diff(f, x));

$$fx := 49$$

- fy := subs(x = 2, y = 3, diff(f, y));

$$fy := 39$$

- N := vector([-fx, -fy, 1]);
  V := vector([x, y, z]);

$$N := [-49 \quad -39 \quad 1]$$

$$V := [x \quad y \quad z]$$

Given the normal to a plane, the equation of the plane is

- q := dotprod(N, V) = d;

$$q := -49 x - 39 y + z = d$$

We find the value of the constant d by demanding that the plane contain the point of contact, here, (2, 3, f(2,3)).

- q1 := subs(z = f, x = 2, y = 3, q);

$$q1 := -144 = d$$

- tanplane := subs(d = -144, q);

$$tanplane := -49\,x - 39\,y + z = -144$$

Example 8

Find a point on the surface $z = x^3 - 2\,x\,y^2$ at which the tangent plane is parallel to the plane $1909\,x - 1428\,y - 170\,z = 510$.

We describe the surface in Maple by entering $z = f(x,y)$ as

- f := x^3 - 2*x*y^2;

$$f := x^3 - 2\,x\,y^2$$

Two planes are parallel if they have the same normal vector, and we can read the components of the normal vector from the given plane.

- N := vector([1909, -1428, -170]);

$$N := [1909 \quad -1428 \quad -170]$$

This normal vector must be proportional to the surface's normal vector, $[-f_x, -f_y, 1]^t$. This proportionality leads to the following three equations.

- e1 := -diff(f,x) = m*N[1];
  e2 := -diff(f,y) = m*N[2];
  e3 :=   1   = m*N[3];

$$e1 := -3\,x^2 + 2\,y^2 = 1909\,m$$

$$e2 := 4\,x\,y = -1428\,m$$

$$e3 := 1 = -170\,m$$

The exact solution to these equations is found by

- q := solve({e1, e2, e3}, {x, y, m});

$$q := \left\{ m = \frac{-1}{170}, x = -\frac{20}{63}\%1^3 + \frac{1909}{1071}\%1, y = \%1 \right\}$$

$$\%1 := \text{RootOf}(-22491 + 3400\_Z^4 + 19090\_Z^2)$$

The compressed form of the solution contained in the **RootOf** structure is expressed as radicals by using the **allvalues** command with the optional parameter 'd' that limits the output to just the distinct solutions.

- q1 := allvalues(q, 'd');

$$q1 := \left\{ m = \frac{-1}{170}, y = \frac{1}{340}\sqrt{\%2}, x = \frac{1}{123807600}\%2^{3/2} + \frac{1909}{364140}\sqrt{\%2} \right\},$$

$$\left\{ m = \frac{-1}{170}, y = -\frac{1}{340}\sqrt{\%2}, x = -\frac{1}{123807600}\%2^{3/2} - \frac{1909}{364140}\sqrt{\%2} \right\},$$

$$\left\{ m = \frac{-1}{170}, y = \frac{1}{340}\sqrt{\%1}, x = \frac{1}{123807600}\%1^{3/2} + \frac{1909}{364140}\sqrt{\%1} \right\},$$

$$\left\{ m = \frac{-1}{170}, y = -\frac{1}{340}\sqrt{\%1}, x = -\frac{1}{123807600}\%1^{3/2} - \frac{1909}{364140}\sqrt{\%1} \right\}$$

$$\%1 := -324530 - 170\sqrt{6703057}$$

$$\%2 := -324530 + 170\sqrt{6703057}$$

The complexity of these solutions is surprising in light of the simplicity of the statement of the problem. Converting the solutions to floating-point form gives insight into the meaning of the expressions in the set q1.

- evalf(q1);

$$\{ m = -.005882352941, y = 1.000019313, x = 2.099959447 \},$$
$$\{ y = -1.000019313, m = -.005882352941, x = -2.099959447 \},$$
$$\{ x = -.816512349\ I, m = -.005882352941, y = 2.571914561\ I \},$$
$$\{ x = .816512349\ I, y = -2.571914561\ I, m = -.005882352941 \}$$

There are two real solutions for the point (x, y), and we exhibit the coordinates for each.

- P1 := subs(q1[1], [x,y]);
  P2 := subs(q1[2], [x,y]);

$$P1 := \left[\frac{1}{123807600}(-324530 + 170\sqrt{6703057})^{3/2} + \frac{1909}{364140}\sqrt{-324530 + 170\sqrt{6703057}}\,, \right.$$
$$\left. \frac{1}{340}\sqrt{-324530 + 170\sqrt{6703057}} \right]$$

$$P2 := \left[-\frac{1}{123807600}(-324530 + 170\sqrt{6703057})^{3/2} \right.$$
$$\left. -\frac{1909}{364140}\sqrt{-324530 + 170\sqrt{6703057}}\,, -\frac{1}{340}\sqrt{-324530 + 170\sqrt{6703057}} \right]$$

Example 9

Should the function f(x,y,z) = 0 define z = z(x,y) implicitly, the partial derivatives ∂z/∂x and ∂z/∂y can then be obtained by a process of implicit differentiation. Illustrate this process for the function f(x,y,z) = 2x² + 3y² + 5z² - 7 = 0.

• f := 2*x^2 + 3*y^2 + 5*z^2 - 7;

$$f := 2\,x^2 + 3\,y^2 + 5\,z^2 - 7$$

To see that this formula defines some surface z = z(x,y) implicitly, obtain a plot of the surface by Maple's **implicitplot3d** command in the plots package.

• implicitplot3d(2*x^2 + 3*y^2 + 5*z^2 - 7, x = -2..2, y = -1.6..1.6,
            z = -1.2..1.2, axes = boxed);

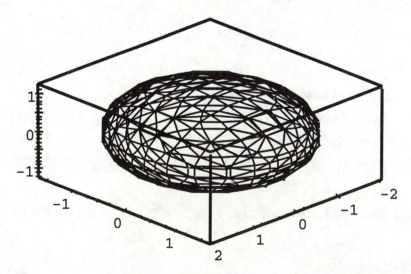

The surface is an ellipsoid for which each cross section is an ellipse. In this example we can actually solve for z = z(x,y) explicitly.

- solve(f, z);

$$\frac{1}{5}\sqrt{-10\,x^2 - 15\,y^2 + 35}\,, -\frac{1}{5}\sqrt{-10\,x^2 - 15\,y^2 + 35}$$

There are two branches to this surface, the top half and the bottom half of the closed surface called an ellipsoid. We next obtain partial derivatives implicitly.

- q := subs(z = z(x,y), f);

$$q := 2\,x^2 + 3\,y^2 + 5\,z(x,y)^2 - 7$$

- q1 := diff(q, x);

$$q1 := 4\,x + 10\,z(x,y)\left(\frac{\partial}{\partial x}z(x,y)\right)$$

- solve(q1, diff(z(x,y),x));

$$-\frac{2}{5}\frac{x}{z(x,y)}$$

- q2 := diff(q, y);

$$q2 := 6\,y + 10\,z(x,y)\left(\frac{\partial}{\partial y}z(x,y)\right)$$

- solve(q2, diff(z(x,y),y));

$$-\frac{3}{5}\frac{y}{z(x,y)}$$

Example 10

Obtain formulas for the implicit derivatives $\partial z/\partial x$ and $\partial z/\partial y$ if z(x,y) is defined implicitly by the function f(x,y,z) = 0.

- f := 'f':

- q := f(x, y, z);

$$q := \mathrm{f}(x,y,z)$$

- q1 := subs(z = z(x,y), q);

$$q1 := \mathrm{f}(x,y,z(x,y))$$

- q2 := diff(q1, x);

$$q2 := D_1(f)(x, y, z(x, y)) + D_3(f)(x, y, z(x, y)) \left(\frac{\partial}{\partial x} z(x, y) \right)$$

- zx := solve(q2, diff(z(x,y), x));

$$zx := - \frac{D_1(f)(x, y, z(x, y))}{D_3(f)(x, y, z(x, y))}$$

This implicit partial derivative is the negative of the ratio of $\partial f/\partial x$ to $\partial f/\partial z$. The numeric subscripts used by Maple are standard mathematical notation for differentiation with respect to the first (in the numerator) and the third (in the denominator) variables, which are here, respectively, x and z. This implicit derivative is most succinctly expressed in subscript notaton via the formula $z_x = - f_x/f_z$.

The implicit partial of z(x,y) with respect to y is also expressed as $z_y = - f_y/f_z$. The Maple computation is left to the reader.

Example 11

Obtain a formula for the normal to a surface defined implicitly by a function of the form f(x,y,z) = 0.

We already know how to form the normal vector for a surface given explicitly as z = g(x,y). In fact, this normal is the vector $[-g_x, -g_y, 1]^t$. Moreover, in Example 10 we saw how to express the implicit partial derivatives $z_x = g_x$, and $z_y = g_y$ in terms of the partial derivatives f_x, f_y, and f_z. Thus, the desired normal to the surface defined implicitly by f(x,y,z) = 0 is

- n := vector([-gx, -gy, 1]);

$$n := [-gx \quad -gy \quad 1]$$

but with gx and gy in n replaced as indicated below.

- N := subs(gx = -fx/fz, gy = -fy/fz, op(n));

$$N := \left[\frac{fx}{fz}, \frac{fy}{fz}, 1 \right]$$

If we multiply N by the quantity fz, we still have a vector in the same direction, namely, normal to the surface defined implicitly by f(x,y,z) = 0. Hence, we take our normal as

- evalm(fz*N);

$$[fx \quad fy \quad fz]$$

This normal vector is so important in the applications that it has a special name, the gradient vector for f(x,y,z). Maple has a **grad** command in its linear algebra package for computing gradients. To apply the **grad** command, you need to include a list of coordinates with respect to which the derivatives will be computed.

- grad(f(x,y,z),   [x,y,z]);

$$\left[\frac{\partial}{\partial x} f(x,y,z) \quad \frac{\partial}{\partial y} f(x,y,z) \quad \frac{\partial}{\partial z} f(x,y,z) \right]$$

Example 12

At point P1, the rate of change of the function f(x,y,z), in the direction from P1 to P2, is called the directional derivative of f. If P1 = (1, 2, -3) and P2 = (5, -7, 4), find the directional derivative at P1 in the direction of P2 if f(x,y,z) = 2x² + 3xy³ - 7yz² + 5y - 3z.

The required rate of change is determined along a straight line connecting P1 to P2. To guarantee a unique parameter along this line, it is taken as arc length. So, if P1 and P2 are interpreted as vectors, then a vector connecting P1 to P2 is **V = P2 - P1**, and the line would be **P1** + s **V** where s is arc length.

- f := 2*x^2 + 3*x*y^3 - 7*y*z^2 + 5*y - 3*z;

$$f := 2\,x^2 + 3\,x\,y^3 - 7\,y\,z^2 + 5\,y - 3\,z$$

- P1 := vector([1, 2, -3]);
  P2 := vector([5, -7, 4]);
  V := evalm(P2 - P1);
  q := s = Int(sqrt(dotprod(V,V)), r = 0..t);

$$P1 := [1 \quad 2 \quad -3]$$

$$P2 := [5 \quad -7 \quad 4]$$

$$V := [4 \quad -9 \quad 7]$$

$$q := s = \int_0^t \sqrt{146}\, dr$$

- t := solve(value(q), t);

$$t := \frac{1}{146} s \sqrt{146}$$

- P1P2 := evalm(P1 + t*V);

$$PIP2 := \left[1 + \frac{2}{73}s\sqrt{146} \quad 2 - \frac{9}{146}s\sqrt{146} \quad -3 + \frac{7}{146}s\sqrt{146} \right]$$

Express f as a function of s along the line from P1 to P2.

- fs := subs(x = P1P2[1], y = P1P2[2], z = P1P2[3], f);

$$fs := 2\left(1 + \frac{2}{73}s\sqrt{146}\right)^2 + 3\left(1 + \frac{2}{73}s\sqrt{146}\right)\left(2 - \frac{9}{146}s\sqrt{146}\right)^3$$
$$- 7\left(2 - \frac{9}{146}s\sqrt{146}\right)\left(-3 + \frac{7}{146}s\sqrt{146}\right)^2 + 19 - \frac{33}{73}s\sqrt{146}$$

Now we can differentiate with respect to arc length. This rate of change, when s = 0, is what we mean by the directional derivative.

- DD := diff(fs, s);

$$DD := \frac{8}{73}\left(1 + \frac{2}{73}s\sqrt{146}\right)\sqrt{146} + \frac{6}{73}\sqrt{146}\left(2 - \frac{9}{146}s\sqrt{146}\right)^3$$
$$- \frac{81}{146}\left(1 + \frac{2}{73}s\sqrt{146}\right)\left(2 - \frac{9}{146}s\sqrt{146}\right)^2\sqrt{146}$$
$$+ \frac{63}{146}\sqrt{146}\left(-3 + \frac{7}{146}s\sqrt{146}\right)^2$$
$$- \frac{49}{73}\left(2 - \frac{9}{146}s\sqrt{146}\right)\left(-3 + \frac{7}{146}s\sqrt{146}\right)\sqrt{146} - \frac{33}{73}\sqrt{146}$$

But we are at point P1 only when s = 0. Hence, the directional derivative we want is

- subs(s = 0, DD);

$$\frac{877}{146}\sqrt{146}$$

Example 13

Repeat Example 12 for a symbolic function f(x,y,z) and generalize to a formula for the directional derivative.

- f := 'f':

- fs := f(P1P2[1], P1P2[2], P1P2[3]);

$$fs := f\left(1 + \frac{2}{73}s\sqrt{146}, 2 - \frac{9}{146}s\sqrt{146}, -3 + \frac{7}{146}s\sqrt{146}\right)$$

- DD := subs(s = 0, diff(fs, s));

$$DD := \frac{2}{73}D_1(f)(1, 2, -3)\sqrt{146} - \frac{9}{146}D_2(f)(1, 2, -3)\sqrt{146} + \frac{7}{146}D_3(f)(1, 2, -3)\sqrt{146}$$

The directional derivative DD has the components of the gradient of f, namely, f_x, f_y, and f_z. In addition, the numbers seem to be the components of the vector **V** from P1 to P2, but normalized. In fact,

- U := normalize(V);

$$U := \left[\frac{2}{73}\sqrt{146} \quad -\frac{9}{146}\sqrt{146} \quad \frac{7}{146}\sqrt{146}\right]$$

When the directional derivative exists, it can be computed as the dot product of the gradient and a unit vector in the direction of interest. Thus,

- dotprod(grad(f(x,y,z), [x, y, z]), U);

$$\frac{2}{73}\left(\frac{\partial}{\partial x}f(x, y, z)\right)\sqrt{146} - \frac{9}{146}\left(\frac{\partial}{\partial y}f(x, y, z)\right)\sqrt{146} + \frac{7}{146}\left(\frac{\partial}{\partial z}f(x, y, z)\right)\sqrt{146}$$

As claimed, the dot product between grad(f) and U reproduces the directional derivative.

Example 14

Obtain a formula for expanding a function z = f(x,y) in a Taylor series, and show that the linear part of such a series is the tangent plane at the point of expansion. This implies that the tangent plane is indeed the local approximation to the function.

- readlib(mtaylor):
- q := mtaylor(f(x,y), [x = a, y = b], 3);

$$q := \mathrm{f}(a,b) + D_1(f)(a,b)\,(x-a) + D_2(f)(a,b)\,(y-b) + \frac{1}{2}D_{1,1}(f)(a,b)\,(x-a)^2$$

$$+ (x-a)\,D_{1,2}(f)(a,b)\,(y-b) + \frac{1}{2}D_{2,2}(f)(a,b)\,(y-b)^2$$

A careful inspection of the result in q shows that the function and all its derivatives are evaluated at the point of expansion, (a, b). The linear part of the series consists of the terms $f(a,b) + f_x(a,b)\,(x - a) + f_y(a,b)\,(y - b)$. We now compare this to the tangent plane whose point of contact with the surface z = f(x,y) is the point (a, b).

The normal to the surface is $[-f_x, -f_y, 1]^t$, and this is the normal to the tangent plane which is therefore of the form

- q1 := dotprod(vector([-fx, -fy, 1]), vector([x, y, z])) = d;

$$q1 := \text{-}fx\,x - fy\,y + z = d$$

To evaluate d requires that (a, b, f(a,b)) be a point on this plane.

- q2 := subs(x = a, y = b, z = f, q1);

$$q2 := \text{-}fx\,a - fy\,b + f = d$$

- q3 := solve(q2, d);

$$q3 := \text{-}fx\,a - fy\,b + f$$

- q4 := subs(d = q3, q1);

$$q4 := \text{-}fx\,x - fy\,y + z = \text{-}fx\,a - fy\,b + f$$

Solving q4 for z = f(x,y) leads to

- q5 := solve(q4, z);

$$q5 := fx\,x + fy\,y - fx\,a - fy\,b + f$$

- q6 := collect(q5, [fx, fy]);

$$q6 := (x - a)\,fx + (y - b)\,fy + f$$

Except for the ordering of terms, and an insertion of (a, b) as the arguments of f and the partial derivatives, this result confirms that f(x,y) is approximated to first order by its tangent plane.

Example 15

Find the minimum distance between the two lines whose parametric equations are

- t := 't':
- xs := -1 + s; ys := 2*s;   zs := 3 + s;
  xt := 3*t;   yt := 2 + t; zt := -1 + 2*t;

$$xs := -1 + s$$

$$ys := 2\,s$$

$$zs := 3 + s$$

$$xt := 3\,t$$

$$yt := 2 + t$$

$$zt := -1 + 2\,t$$

The distance between these two lines is

- q := sqrt((xs - xt)^2 + (ys - yt)^2 + (zs - zt)^2);

$$q := \sqrt{21 - 2\,s - 6\,t + 6\,s^2 - 14\,s\,t + 14\,t^2}$$

The equations yielding a candidate for a minimum are $\partial q/\partial s = 0$ and $\partial q/\partial t = 0$.

- q1 := solve({diff(q, s) = 0, diff(q, t) = 0}, {s, t});

$$q1 := \left\{ s = 1, t = \frac{5}{7} \right\}$$

- d1 := subs(q1, q);

$$d1 := \frac{1}{7}\sqrt{125}\,\sqrt{7}$$

Example 16

On the surface given implicitly by the function $z^2 = x^2 y + 4$, find the point which is closest to the origin.

The quantity to be minimized is the distance from the origin, but the surface on which the point must lie is a constraint. Hence, this is a constrained optimization problem and, as such, is a candidate for the Lagrange multiplier technique. We choose to minimize the square of the distance to the origin in order to avoid the square root in all our equations. Our Lagrangian will be F = f - m g, with m being the Lagrange multiplier. The objective function f, the constraint function g, and the Lagrangian function F are entered as

- f := x^2 + y^2 + z^2;
  g := z^2 - x^2*y - 4;
  F := f - m*g;

$$f := x^2 + y^2 + z^2$$

$$g := z^2 - x^2 y - 4$$

$$F := x^2 + y^2 + z^2 - m\left(z^2 - x^2 y - 4\right)$$

The governing equations of the Lagrange multiplier method are then

- e1 := diff(F, x) = 0;
  e2 := diff(F, y) = 0;
  e3 := diff(F, z) = 0;
  e4 := diff(F, m) = 0;

$$e1 := 2x + 2mxy = 0$$

$$e2 := 2y + mx^2 = 0$$

$$e3 := 2z - 2mz = 0$$

$$e4 := -z^2 + x^2 y + 4 = 0$$

An exact solution of these equations is found via

- q := solve({e1, e2, e3, e4}, {x, y, z, m});

$$q := \{x = 0, y = 0, z = 2, m = 1\}, \{z = -2, x = 0, y = 0, m = 1\}, \left\{ z = 0, \right.$$

$$\left. y = -\frac{1}{8}\operatorname{RootOf}(\_Z^6 - 32)^4, x = \operatorname{RootOf}(\_Z^6 - 32), m = \frac{1}{4}\operatorname{RootOf}(\_Z^6 - 32)^2 \right\},$$

$$\{x = \operatorname{RootOf}(-2 + \_Z^2), y = -1, z = \sqrt{2}, m = 1\},$$

$$\{x = \operatorname{RootOf}(-2 + \_Z^2), y = -1, z = -\sqrt{2}, m = 1\}$$

The values of the objective function f, computed at each of the first two solutions found above, are

- sol1 := subs(q[1], f);
  sol2 := subs(q[2], f);

$$sol1 := 4$$

$$sol2 := 4$$

The third exact solution contains a **RootOf** which is converted to radicals via

- q3 := allvalues(q[3], 'd');

$$q3 := \left\{ z = 0, y = -2^{1/3}, m = \frac{1}{2}2^{2/3}, x = 2^{5/6} \right\}, \left\{ \right.$$

$$z = 0, m = \frac{1}{2}\left(\frac{1}{2}+\frac{1}{2}I\sqrt{3}\right)^2 2^{2/3}, x = \left(\frac{1}{2}+\frac{1}{2}I\sqrt{3}\right)2^{5/6}, y = -\left(\frac{1}{2}+\frac{1}{2}I\sqrt{3}\right)^4 2^{1/3} \right\},$$

$$\left\{ z = 0, y = -\left(-\frac{1}{2}+\frac{1}{2}I\sqrt{3}\right)^4 2^{1/3}, m = \frac{1}{2}\left(-\frac{1}{2}+\frac{1}{2}I\sqrt{3}\right)^2 2^{2/3}, \right.$$

$$x = \left(-\frac{1}{2}+\frac{1}{2}I\sqrt{3}\right)2^{5/6} \right\}, \left\{ z = 0, y = -2^{1/3}, m = \frac{1}{2}2^{2/3}, x = -2^{5/6} \right\}, \left\{ z = 0, \right.$$

$$y = -\left(-\frac{1}{2}-\frac{1}{2}I\sqrt{3}\right)^4 2^{1/3}, m = \frac{1}{2}\left(-\frac{1}{2}-\frac{1}{2}I\sqrt{3}\right)^2 2^{2/3}, x = \left(-\frac{1}{2}-\frac{1}{2}I\sqrt{3}\right)2^{5/6} \right\}, \left\{ \right.$$

$$z = 0, x = \left(\frac{1}{2}-\frac{1}{2}I\sqrt{3}\right)2^{5/6}, y = -\left(\frac{1}{2}-\frac{1}{2}I\sqrt{3}\right)^4 2^{1/3}, m = \frac{1}{2}\left(\frac{1}{2}-\frac{1}{2}I\sqrt{3}\right)^2 2^{2/3} \right\}$$

The optional argument "d" in the **allvalues** command prevents spurious pairs from being listed as solutions. However, there should be two real solutions in this group instead of just one. Unfortunately, Maple has dropped the solution $(2^{5/6}, -2^{1/3}, 0)$, which is companion to q3[4].

- sol3 := subs(q3[4], f);

$$sol3 := 3\,2^{2/3}$$

- evalf(sol3);

$$4.762203156$$

The missing solution will generate the same value for f as sol3. Since this value is already greater than 4, neither is a candidate for the minimum. The fourth and fifth exact solutions are expresed as radicals via

- q4 := allvalues(q[4], 'd');
  q5 := allvalues(q[5], 'd');

$$q4 := \{ x = \sqrt{2}, y = -1, z = \sqrt{2}, m = 1 \}, \{ x = -\sqrt{2}, y = -1, z = \sqrt{2}, m = 1 \}$$

$$q5 := \{ x = \sqrt{2}, y = -1, z = -\sqrt{2}, m = 1 \}, \{ x = -\sqrt{2}, y = -1, z = -\sqrt{2}, m = 1 \}$$

The values of the objective function f at these additional points are found by

- sol5 := subs(q4[1], f);
  sol6 := subs(q4[2], f);
  sol7 := subs(q5[1], f);
  sol8 := subs(q5[2], f);

$$sol5 := 5$$

$$sol6 := 5$$

$$sol7 := 5$$

$$sol8 := 5$$

The minimum distance of 4 occurs at (0, 0, 2) and (0, 0, -2).

It is tempting to dismiss the complexity of this Lagrange multiplier solution as unnecessary, since z^2 in the objective function f can be replaced by $x^2y + 4$ from the constraint function g. This gives an unconstrained optimization in just two variables that is far simpler to solve than the original constrained optimization. In Project 15 we explore the consequences of this naivete.

Exercises

1. Determine the largest possible domain for each function f(x,y). Obtain a plot of this domain in the xy-plane. Plot the surface z = f(x,y) over a representative portion of that domain.

    (a) $f(x,y) = \sqrt{y - x^2 + 4x - 3}$ **(c)** $f(x,y) = \sqrt{e^{-\frac{x}{2}} \cos x - y}$

    (b) $f(x,y) = \sqrt{x^2 + 4y^2 - 4}$ **(d)** $f(x,y) = \sqrt{x^2 - y^2 - 1.9x - 2.1y - 1.7}$.

2. For $f(x, y) = \dfrac{x^2 y}{x^3 + y^3}$, find (if possible)

    (a) f(1.1, 2.1)   **(b)** f(1.1, 1.9)   **(c)** f(0.9, 2)

    (d) f(1, 2)   **(e)** $\lim\limits_{(x,y) \to (1,2)} f(x,y)$.

3. For $f(x,y) = \dfrac{x^2 y}{x^3 + y^3}$, find (if possible)

    (a) f(0.1, 0.1)   **(b)** f(0.1, -0.1)   **(c)** f(-0.1, 0)

    (d) f(0, 0)   **(e)** $\lim\limits_{(x,y) \to (0,0)} f(x,y)$.

4. For $f(x,y) = \tan^{-1}\left(\dfrac{x^2 + y^2 - 1}{x^2 + y^2}\right)$, find (if possible)

    (a) f(0.1, 0.1)   **(b)** f(0.1, -0.1)   **(c)** f(0, 0.1)

    (d) f(0, 0)   **(e)** $\lim\limits_{(x,y) \to (0,0)} f(x,y)$.

5. Plot f(x,y) on an appropriate domain. Compare rectangular and parametric plots for some of these.

    (a) $f(x,y) = x^2 + \frac{1}{16} y^2$ **(c)** $f(x,y) = xe^{-xy}$

    (b) $f(x,y) = 4 - x^2 - y^2$ **(d)** $f(x,y) = e^{-(x^2 + \frac{1}{4}y^2)} \cos x$.

6. For $z = x^3 + y^2 + \sin x$, find

   (a) z_x **(b)** z_y **(c)** z_{xy} **(d)** z_{yx}.

7. For $f(x,y) = x^3 y^2$, find

   (a) $\dfrac{\partial f}{\partial x}$ **(b)** $\dfrac{\partial f}{\partial y}$ **(c)** $\dfrac{\partial^2 f}{\partial x^2}$ **(d)** $\dfrac{\partial^2 f}{\partial y \partial x}$.

8. Define $f(x,y) = e^{xy} \cos y$. Plot $f(x,y)$ for $0 \le y \le 1$, $0 \le y \le 1.5$, and in the same window show, on this surface, a tangent line

   (a) whose slope is given by $f_x(0, 1)$ **(b)** whose slope is given by $f_y(0, 1)$.

9. Define $f(x,y) = x + 2 x y + y^2 - 2$ and take $(x_0, y_0) = (1, 3)$. Find

   (a) $\dfrac{f(x_0 + h, y_0) - f(x_0, y_0)}{h}$ with $h = 0.5$, then $h = 0.1$

   (b) $f_x(x_0, y_0)$

   (c) $\dfrac{f(x_0, y_0 + k) - f(x_0, y_0)}{k}$ with $k = 0.2$, then $k = 0.05$

   (d) $f_y(x_0, y_0)$

   (e) Compare results in parts **(a)**, **(b)**, and then in parts **(c)**, **(d)**.

10. Define $f(x,y) = x \sin y$ and take $(x_0, y_0) = (2, 0.4)$. Find

   (a) $\dfrac{f(x_0 + h, y_0) - f(x_0, y_0)}{h}$ with $h = 0.3$, then $h = 0.05$

   (b) $f_x(x_0, y_0)$

   (c) $\dfrac{f(x_0, y_0 + k) - f(x_0, y_0)}{k}$ with $k = -0.2$, then $k = -0.05$

   (d) $f_y(x_0, y_0)$

   (e) Compare results in parts **(a)**, **(b)**, and then in parts **(c)**, **(d)**.

11. **(a)** Define a surface by $z = 2x^2 + y^2$. Compare a rectangular plot with a parametric plot. In the parametric plot use $x = \sqrt{2}\ r\cos t$, $y = r\sin t$.

    (b) At $(2, 3)$, how fast is z changing in the direction $[0, 1]^t$?, $[1, 0]^t$?, $[1, 2]^t$?

12. **(a)** Plot the <u>surface</u> $f(x, y) = \sqrt{9 - x^2 - y^2}$. Will rectangular or parametric form give a better plot?

    (b) Plot the <u>curve</u> $\mathbf{r}(t) = [\cos t, 3\sin t, \frac{4t}{t + 0.2}]^t$. Show two different views.

13. A point moves along the intersection of $f(x,y) = e^{-x}\cos y$ and the plane $y = \frac{\pi}{3}$. At what rate is $f(x,y)$ changing at the point:

    (a) $(2, \frac{\pi}{3}, 0.07)$ **(c)** $(-1, \frac{\pi}{3}, 1.36)$?

    (b) $(0.5, \frac{\pi}{3}, 0.30)$

14. Define $f(x,y) = x^2 + 3xy$. Find Δf when

    (a) $(x_0, y_0) = (2, 1)$, $\Delta x = 0.01$, and $\Delta y = 0.02$

    (b) $(x_0, y_0) = (0.1, 0.2)$, $\Delta x = 0.2$, and $\Delta y = 0.1$.

15. Define $z = x^2 + 3xy$, $x = 3uv$, and $y = v^3$. Use the Chain Rule to find z_u and z_v; then check your answer by substituting for x and y and finding the partial derivatives directly.

16. **(a)** At the point where $x = \frac{\pi}{3}$ and $y = \frac{1}{3}$, find an equation of the plane tangent to the surface $z = e^{-\sqrt{x^2 + \frac{1}{2}y^2}}\cos x$.

    (b) In the same window, plot the surface and the tangent plane.

17. Find the direction in which $f(x,y) = x^3 - 2xy^2$ increases most rapidly at P. Then find the rate of change of f in that direction.

    (a) $P = (1, 0)$ **(b)** $P = (1, 1)$ **(c)** $P = (-1, 2)$.

18. Define $f(x,y) = x^3 - 2xy^2$.

    (a) Plot $f(x,y)$ on $-2 \le x \le 2$, $-2 \le y \le 4$.

(b) Plot several level curves of f(x,y) for $-2 \leq x \leq 2$, $-2 \leq y \leq 4$.

19. Find the direction in which $f(x,y) = \cos(x + 2y)e^{-\frac{y}{2}}$ increases most rapidly at P. Then find the rate of change in that direction.

   (a) $P = (6, 3)$ (c) $P = (0, -1)$.

   (b) $P = (6, -2)$

20. Define $f(x,y) = e^{-\frac{y}{2}}\cos(x + y)$.

   (a) Plot $f(x,y)$ on $-2.5 \leq x \leq 6.5$, $-2 \leq y \leq 3$.

   (b) Plot several level curves of $f(x,y)$ for $-2.5 \leq x \leq 6.5$, $-2 \leq y \leq 3$.

21. At the point $(9, 12, z(9,12))$ on the surface $z = 5000e^{-(\frac{1}{50}x^2 + \frac{1}{150}y^2)}$, find a direction $\mathbf{u} = [\cos\theta, \sin\theta]^t$ in which the rate of change of z is

   (a) -23.55 (b) 94.32.

22. Find the point of intersection of the line $\mathbf{r}(t) = [1 + 2t, 1 + 0.9t, -1 + 4.1t]^t$ and the surface $z = x^3 - 2xy^2$. At the point of intersection, find the direction in which the rate of change of z is maximum. Then find the rate of change in that direction.

23. Find and classify all relative maxima and minima of f(x, y). Then plot f(x, y) over a region that includes the relative extrema.

   (a) $f(x,y) = xy + x + y - \ln(xy^3)$ (d) $f(x,y) = e^{-\frac{1}{5}\sqrt{x^2 + y^2}}\cos y$

   (b) $f(x,y) = xy + x^2 + y^2 - 2\ln(xy)$ (e) $f(x,y) = xy - x^5 + 5x^3 - 4x - y^2$.

   (c) $f(x,y) = e^{-\sqrt{x^2 + \frac{4}{9}y^2}}\cos x \, \cos y$

24. Find the distance between the point and the plane.

   (a) $(2, 5, 4)$ and $x - y + 2z = 1$ (b) $(3, 3, 9)$ and $2x + y - z = 2$.

25. Find the distance between $\mathbf{r}_1(s)$ and $\mathbf{r}_2(t)$. Display $\mathbf{r}_1(s)$ and $\mathbf{r}_2(t)$ in the same window.

   (a) $\mathbf{r}_1(s) = [0,1,1]^t + s[1,2,1]^t$, $\mathbf{r}_2(t) = [0,1,1]^t + t[1,2,1]^t$

(**b**) $r_1(s) = [1, 1, 1]^t + s[3, -2, 1]^t$, $r_2(t) = [2, -3, 1]^t + t[1, 2, 3]^t$.

26. Find the distance between the point and the surface. Display both in the same window. (See the Point Style option in the Style field.)

(**a**) $(1, 2, 1)$ and $z = x^2 + 2 y^2 + 4$ (**c**) $(1.5, -0.3, 0)$ and $z = e^{-\frac{y}{2}} \cos(x + 2 y)$.

(**b**) $(3, 1, 1.5)$ and $z = x^3 - 2 x y^2$

27. Find the maximum value of z on the disk $x^2 + y^2 \leq 1$.

(**a**) $z = x^3 + 2 x y - y^3$ (**b**) $z = x^3 + 3 x^2 y^2 - y^3$.

28. Find and classify all relative extrema of z. Plot the surface over a region which contains the relative extrema.

(**a**) $z = \dfrac{x^3 - 3 x^2}{1 + y^2}$ (**b**) $z = \dfrac{y^3 - 6 y^2 + 9 y}{1 + x^2}$.

29. Define a function $f(x, y)$ which has a relative maximum at $(2, 3)$. (<u>Hint</u>: The reciprocal of $g(x,y) = (x - 2)^2 + (y - 3)^2$ gives you one approach.) Plot $f(x,y)$.

30. The temperature in degrees Celsius at each point (x,y) in the region $0 \leq x \leq 1, 0 \leq y \leq 1$, is given by $T(x,y) = 0.1 x y - x^3 - y^3 + x$. Find the maximum and minimum temperatures in this region.

31. A box with a rectangular base and no top is to hold 500 cubic centimeters. Find the dimensions of the box for which the surface area (the base and four sides) is minimum.

32. Use Lagrange multipliers to find the extreme values of $f(x,y,z) = x - 2 y + 3 z$, subject to the constraint $\dfrac{x^2 + y^2}{4 + z^2} = 1$.

Projects

1. Find a point P on the surface $f(x, y) = \frac{1}{2} x y^2$ where the tangent plane is parallel to the plane $3 x + 2 y - 6 z + 18 = 0$. In the same window plot the surface $f(x,y)$, the upward normal to $f(x,y)$ at P, and the plane $3 x + 2 y - 6 z + 18 = 0$.

2.   Repeat the directions in Problem 1 for the surface $f(x,y) = e^{xy}\cos y$ and the plane $2.088 x - 2.278 y - 4.755 z - 1 = 0$.

3.   The surface of a mountain is defined by $z(x,y) = 25 - x^2 - y^2$, where the direction along the positive y-axis is north. $P = (3, 2, 12)$ is a point on the side of the mountain.

If you start walking from point P, at what angle does the trail slope if you walk

   (a) north   **(b)** east   **(c)** southwest?

   (d) Are you gaining or losing elevation in each case **(a)**, **(b)**, **(c)**?

   (e) Looming in the distance is a taller mountain ($w(x,y)$) whose summit base is 10 units north and 20 units east of the base of the first summit. The elevation of the summit is 32 units. At the point $(21, 7, w(21, 7))$ on the side of this taller mountain, the trail has an upward slope of 1 in a due west direction. At the same point the trail has a downward slope of 8 in a due south direction. Find a function $w(x, y)$ that describes the surface of this mountain.

4.   A pipe running parallel to the ground rests on a support that has a square base and a top surface with parabolic cross sections. The four corners of the top surface are at $(0, \pm2, 1)$, and $(4, \pm2, 1)$. The vertex of each cross section is on the x-axis, and the corners of the base are at $(0, \pm2, -3)$, and $(4, \pm2, -3)$. Generate a picture of the support and show two different views.

5.   Define $f(x, y) = 4 x^2 e^y - 2 x^4 - e^{4y}$.

   (a) Find the critical points of $f(x,y)$.

   (b) Classify each critical point (x_0, y_0) by comparing $f(x_0,y_0)$ with each $f(x_k,y_k)$ where the (x_k, y_k) are points on a circle of small radius having center at (x_0, y_0). If this small circle of fixed radius is expressed in polar coordinates, the values of $f(x,y)$ above that circle can be given in terms of the parameter on the circle. Thus, a table of function values would reveal whether or not all of the values at the selected points on the circle were greater than or less than the value at the critical point.

   (c) Check your answers in part **(b)** by applying the Second Partials Test to each critical point.

6.   **(a)** Define a function $f(x,y)$ that has a relative maximum value at $(x, y) = (2, 1)$. Plot your function.

   (b) Define a function $f(x,y)$ that has critical points at $(x, y) = (\pm1, 0)$. Classify each critical point. Plot your function.

(c) Define a function f(x,y) that has a relative maximum at (1, 0) and a relative minimum at (2, 0). Plot your function.

7.   Write Maple instructions that will plot level curves of $z = \sqrt{x^2 + \frac{y^2}{4} + 1}$. Be sure to allow for a varying number of contours, and use Maple's **implicitplot** command to draw individual contours.

8.   Plot each surface and then generate contour plots and cross sections as directed. Reconcile the different types of plots of each surface.

(a) $z = 4xy - x^4 - y^4$
    contourplots at z = 0, 0.5, 1.0, 1.5, 1.9, 1.98
    cross sections for x = 0.1, 0.5, 1.0, 1.5
    cross sections for y = x - 0.5, y = x, y = x + 0.5

(b) $z = \sin(\sqrt{x^2 + y^2})$
    contourplots for r = 0.1, 0.5, 1.0, $\frac{\pi}{2}$, 1.7

(c) $z = 3xe^y - x^3 - e^{3y}$
    contourplots at z = -1, 0, 0.5, 0.9, 1.1, 1.5
    cross sections for x = 0, 0.3, 0.6, 1.0, 1.2
    cross sections for y = x-1, y = x, y = x + 1

9.   The nose cone of a space probe has equation $z = -\sqrt{16 - 4x^2 - 4y^2}$. When the probe reenters the atmosphere, the surface heats according to the function
    $T(x, y\ z) = 8y^2 + 4xz - 16x - 1200$. Find the hottest point on the probe's surface.

10. Define $f(x,y) = \sin(x^2 + y^2)$.

(a) Find the multivariate Taylor series expansion of f(x, y) about the origin, with n = 8. Call it t(x,y). (See **?mtaylor**)

(b) Compute and list the values of $|f(x,y) - t(x,y)|$ for x = 0.2, 0.4, 0.6, y = 0.5, 0.8, 1.1.

(c) In the same window, plot f(x,y) and t(x,y) for $0 \le x \le 1$, $0.5 \le y \le 1.5$.

11. Define $f(x,y) = e^y \cos x$.

(a) Find the multivariate Taylor series expansion of f(x,y) about the origin, with n = 4. Call it t(x,y).

(b) Compute and list the values of $\left| f(x,y) - t(x,y) \right|$ for x = 0.2, 0.4, 0.6, 0.8, y = 0.5, 0.8, 1.1.

(c) In the same window, plot f(x,y) and t(x,y) for $0 \le x \le 1$, $0 \le y \le 2$.

12. Define mt(n) to be the nth order multivariate Taylor polynomial about (0, 0) of

$f(x,y) = e^y * \cos(x)$. Make a movie in which each frame shows a plot of f(x,y) and mt(k) for $0 \le x \le \pi, 0 \le y \le 1$, as k takes on the values 2..5.

13. The size of box that can be mailed by parcel post is limited by the total of length and girth, while for air baggage, box size is limited by the total of the three dimensions. Assume that, in this case, these limits are:

$$L + 2W + 2D \le 84 \text{ inches}$$
$$L + W + D \le 54 \text{ inches}$$

A company wishes to have a box, of maximum allowable volume, that it can ship by either method. What are the dimensions of the box it should order?

-Technology Review (The Bent)

14. A tetherball is constrained to move in three dimensions at a constant distance R from a fixed point P, which is at the top of a vertical pole. The ball may be considered a point mass attached to a frictionless, weightless cord. The angular displacement below the horizontal plane through P is denoted by β. Initially, with $\beta = 0$, the ball is given a tangential, horizontal velocity of $(2gR)^{1/2}$. What maximum β is attained when the ball descends to the next low point of the trajectory, which occurs before there is significant wrapping around the pole?

-R. Wilson Rowland (The Bent)

15. Explore the capabilities of the following Maple commands (in the plots package): **densityplot, gradplot, gradplot3d, surfdata, tubeplot.**

16. Solve the constrained optimization problem of Example 6 by replacing z^2 in the objective function f(x,y,z) with z^2 defined by the constraint g(x,y,z) = 0. Solve the resulting unconstrained optimization problem in two variables. Note that you did not get all the solutions obtained in Example 6 via the Lagrange multiplier technique. This is because the constraint g(x,y,z) = 0 does not define z = z(x,y) implicitly along z = 0. Hence, when z = 0, the method of substitution will not yield the extreme value corresponding to z = 0.

Chapter 17

Multiple Integrals

New Maple Commands for Chapter 17

plot3d([f(x,y),g(x,y),h(x,y)], x=a..b,y=c..d);	plot of a surface given parametrically
simplify(expression,symbolic);	simplifies expression, letting $\sqrt{x^2}$ -> x
v.(1..6);	creates the sequence v1,v2,v3,v4,v5,v6
with(linalg);	loads the linear algebra package
matrix(n,m,(i,j)->f(i,j));	creates n x m matrix with (i,j) entry = f(i,j)
with(plots);	loads the plots package
matrixplot(matrix, heights=histogram);	plots 3d histogram of values in the matrix
with(student);	loads the student package
Doubleint(f(x,y),x,y,R);	creates inert double integral $\iint_R f(x,y)\,dxdy$
Tripleint(f(x,y,z),x,y,z,R);	creates inert triple integral $\iiint_R f(x,y,z)\,dxdydz$

Introduction

Chapter 17 explores integration for functions of several variables. In particular, the notions of double and triple integrals, and of iterated integrals, are encountered. Multiple integrals can represent areas, volumes, centers of mass, moments of inertia, and other physical quantities. But the definitions of multiple integrals parallel the definition of the definite integral seen earlier: the limit of a sum of more and more of smaller and smaller quantities.

Examples

Example 1

Compute the double integral $\int_R (4 + xy)\,dA$, where R = {(x,y) | $0 \le x \le 2$, $-1 \le y \le 1$}, as the limit of an appropriate approximating sum.

First, we draw a figure that illustrates how this double integral represents the volume under the surface z = 4 + xy.

- with(linalg):
  with(plots):
  with(student):

- f := 4 + x*y;

$$f := 4 + x\,y$$

• matrixplot(matrix(6,6,(j,k)->4+(j/3)*(k/3-1)),   heights=histogram,   axes=   boxed);

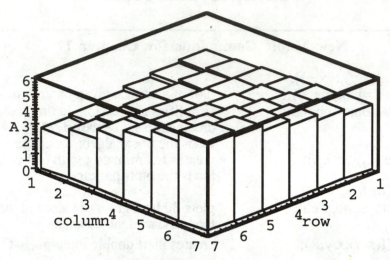

The double integral is the limiting sum of the volumes of the "blocks" shown in the figure. To obtain a representative sum, we need to partition R with a rectangular grid containing N x M rectangles. The evaluation points will correspond to "left side" in both the x and y directions. As N and M become infinite, the volumes of the "blocks" become smaller, and there are more and more blocks. In the limit, this summation becomes the volume under the surface f(x,y).

• dx := 2/N:
• dy := 2/M:

• xj := j*dx;
  yk := -1 + k*dy;
  fjk := subs(x = xj, y = yk, f);
  v := Sum(Sum(fjk * dx*dy, j = 0..N-1), k = 0..M-1);
  V := Limit(Limit(v, N = infinity), M = infinity);
  vol := value(V);

$$xj := 2\,\frac{j}{N}$$

$$yk := -1 + 2\,\frac{k}{M}$$

$$fjk := 4 + 2\,\frac{j\left(-1 + 2\,\dfrac{k}{M}\right)}{N}$$

$$v := \sum_{k=0}^{M-1} \left(\sum_{j=0}^{N-1} \left(4 \frac{4 + 2\dfrac{j\left(-1 + 2\dfrac{k}{M}\right)}{N}}{NM} \right) \right)$$

$$V := \lim_{M \to \infty} \lim_{N \to \infty} \sum_{k=0}^{M-1} \left(\sum_{j=0}^{N-1} \left(4 \frac{4 + 2\dfrac{j\left(-1 + 2\dfrac{k}{M}\right)}{N}}{NM} \right) \right)$$

$$vol := 16$$

The reader who imitates this computation using the symbol xi instead of xj will note that Maple interprets "xi" as the name of a Greek letter and prints that Greek letter to the screen. On the other hand, the reader may want to use the **leftsum** command in the student package to create the sums needed in this demonstration.

• vv := leftsum(leftsum(f, x = 0..2, N), y = -1..1, M);

$$vv := 2 \frac{\displaystyle\sum_{i=0}^{M-1} \left(2 \dfrac{\displaystyle\sum_{i=0}^{N-1} \left(4 + 2\dfrac{i\left(-1 + 2\dfrac{i}{M}\right)}{N} \right)}{N} \right)}{M}$$

• limit(limit(value(vv), N = infinity), M = infinity);

16

Example 2

Evaluate as an iterated integral the double integral in Example 1.

Maple's student package contains a **Doubleint** command that writes, as an inert integral, the notation for a double integral.

• Doubleint(f, x, y, R);

$$\iint\limits_R 4 + x\,y\,dx\,dy$$

It is unfortunate that Maple omits the parentheses around the integrand. In addition, the symbol Maple uses, and mathematicians often use, for the double integral contains dx dy, leading to a confusion between the double integral and its iterated counterpart.

• q := Doubleint(f, x = 0..2, y = -1..1);

$$q := \int_{-1}^{1}\int_{0}^{2} 4 + x\,y\,dx\,dy$$

It is not clear whether Maple intends the symbol $\iint_R f(x,y)\,dxdy$ to represent a double integral or an iterated integral. However, it does appear safe to declare that q is an iterated integral whose value can be obtained by using Maple's **value** command.

• value(q);

$$16$$

For the purist, we illustrate the use of iterated **Int** and **int** commands to evaluate the iterated integral.

• Int(Int(f,x = 0..2), y = -1..1) = int(int(f, x = 0..2), y = -1..1);

$$\int_{-1}^{1}\int_{0}^{2} 4 + x\,y\,dx\,dy = 16$$

Example 3

Find the volume under the surface $z = x^2y$ and above the region R that is bounded by the lines $x = 3$ and $y = x/3$, and the curve $y = x(4-x)$.

First, a diagram of the region R.

- p1 := plot({[3, x, x = 0..5], [x, x/3, x = 0..4], [x, x*(4-x), x = 0..4]}):

- p2 := textplot([1.5, 2, `R`]):

- display([p1,p2]);

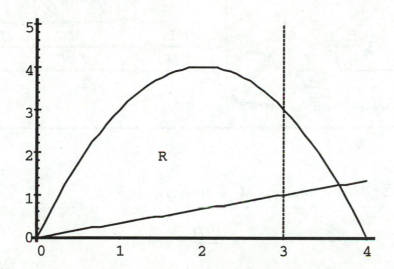

The region R lies beneath the parabola, above the oblique line, and to the left of the vertical line.

To reinforce the notion of the double integral as a limit of a sum, we use the **limit** and **rightsum** commands to implement the definition of the double integral.

- v := rightsum(rightsum(x^2*y, y = x/3..x*(4-x), N), x = 0..3, M);

$$v := 3 \left. \left(\frac{\left(3\dfrac{i\left(4-3\dfrac{i}{M}\right)}{M} - \dfrac{i}{M}\right)\left(\displaystyle\sum_{i=1}^{N}\left(9\dfrac{i^2\left(\dfrac{i}{M} + \dfrac{i\left(3\dfrac{i\left(4-3\dfrac{i}{M}\right)}{M} - \dfrac{i}{M}\right)}{N}\right)}{M^2}\right)\right)}{N}\right) \middle/ M \right.$$

$$\sum_{i=1}^{M}$$

- limit(limit(value(v), N = infinity), M = infinity);

$$\frac{1971}{35}$$

We check this computation by a direct evaluation of the appropriate iterated integral.

- q := Doubleint(x^2*y, y = x/3..x*(4-x), x = 0..3);

$$q := \int_0^3 \int_{\frac{1}{3}x}^{x(4-x)} x^2\, y\, dy\, dx$$

- value(q);

$$\frac{1971}{35}$$

Example 4

Find the volume under the surface z = xy and above the first quadrant region R that is outside the circle r = 2 but inside the cardioid r = 3(1 + cos(t)).

This volume can be found by a double integral in polar coordinates, provided the surface is expressed as z = z(r,t), with t being the polar angle. Since x = r cos(t) and y = r sin(t), the surface is z = r² sin(t) cos(t). It also helps to have a diagram of the region over which the integration is to take place. The figure is constructed in stages. First, the two bounding curves are sketched, then the region R is retraced with thickened boundaries, and finally the two components of the final figure are displayed together. We also included a sketch of a representative "wedge" that suggests the element of area in polar coordinates.

- p1 := plot({[2, t, t = 0..Pi/2], [3*(1 + cos(t)), t, t = 0..Pi/2], [t, 0, t = 2..6],
         [t, Pi/2, t = 2..3]}, coords = polar, scaling = constrained, thickness = 3):

- p2 := plot({[2, t, t = 0..2*Pi], [3*(1 + cos(t)), t, t = 0..2*Pi], [t, Pi/4, t = 0..5.12],
         [t, Pi/4 + .1, t = 0..4.899]}, coords = polar, scaling = constrained):

- display([p1, p2]);

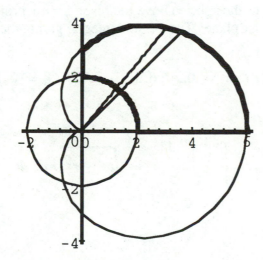

Since the element of area in polar coordinates is r dr dt, we see that the iterated integral giving the required volume is

- q := Int(Int(r^2*sin(t)*cos(t)*r, r = 2..3*(1 + cos(t))), t = 0..Pi/2);

$$q := \int_0^{\frac{1}{2}\pi} \int_2^{3 + 3\cos(t)} r^3 \sin(t) \cos(t) \, dr \, dt$$

• value(q);

$$\frac{3403}{40}$$

Example 5

Find the volume of the "north polar cap" sliced off a sphere of radius a by the plane $z = a\,v$, where $0 < v < 1$.

The sphere is given in cartesian coordinates by $x^2 + y^2 + z^2 = a^2$ so that for $z = a\,v$ we have $x^2 + y^2 = a^2(1 - v^2)$. In cylindrical coordinates, the upper hemisphere is given by $z = \sqrt{(a^2 - r^2)}$ and the portion of the sphere above the plane $z = a\,v$ projects to the interior of the circle $r = a\sqrt{(1 - v^2)}$ in the polar plane. The iterated integral giving the required volume is therefore

• q := Int(Int(sqrt(a^2-r^2)*r, r = 0..a*sqrt(1 - v^2)), t = 0..2*Pi);

$$q := \int_0^{2\pi} \int_0^{a\sqrt{1-v^2}} \sqrt{a^2 - r^2}\, r \, dr \, dt$$

• q1 := value(q);

$$q1 := -\frac{2}{3}\left((a^2 v^2)^{3/2} - (a^2)^{3/2}\right)\pi$$

• simplify(q1, symbolic);

$$-\frac{2}{3}a^3 (v^3 - 1)\,\pi$$

Notice that as v -> 0 (so the polar cap becomes the whole upper hemisphere) the volume becomes $4\pi a^3/3$, the volume of a hemisphere of radius a.

Example 6

Find, by triple integration, the first octant volume bounded by the plane y = x and the cylinder $x^2 + z^2 = 1$.

First, it is useful to generate a plot. Since the plane y = x is not of the form z = z(x,y), it must be plotted parametrically. That induces us to plot the cylinder, which is parallel to the y-axis, parametrically also.

- plot3d({[x, x, free], [x, free, sqrt(1-x^2)]}, x = 0..3, free = 0..3,
                    axes = framed, scaling = constrained);

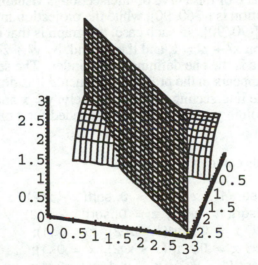

The volume in question is the section of the cylinder that is "in front" of the plane y = x. If the first octant portion of the cylinder is imagined to be a piece of quarter-round molding sitting in a miter-box, and the plane y = x is imagined to be the miter-saw cutting a 45-degree miter, then the volume to be computed is that of the scrap wasted as the quarter-round is trimmed.

There are six possible iterated triple integrals that will yield the required volume. Four of these are fairly obvious, but two depend on understanding (and perhaps visualizing) that the bounding curves in the coordinate planes are the projections of the space curve of intersection of the cylinder and the plane y = x. The space curve of intersection can be sketched with Maple's **spacecurve** command in the plots package after noting that this curve is given parametrically by x(t) = t, y(t) = t, z(t) = $\sqrt{(1 - t^2)}$.

- spacecurve([t, t, sqrt(1 - t^2)], t = 0..1, labels = [x, y, z],   scaling = constrained,
                    axes = normal, orientation = [-347, 70]);

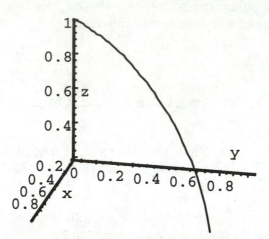

The projection into the yz-plane of this curve of intersection is visualized by rotating the frame of the plot so that the orientation is [-360, 90], while the projection into the xz-plane is seen by rotating to an orientation of [-90, 90]. In each case, the graph is that of a quarter circle, the first governed by the equation $x^2 + z^2 = 1$, and the second, by $y^2 + z^2 = 1$. The first projection is intuitive, since that formula is the one defining the cylinder. The second formula is the challenge, since it nowhere appears in the problem statement. It is determined by eliminating from the two equations of the intersecting surfaces, namely, $y = x$ and $x^2 + z^2 = 1$, the variable x, thereby creating an equation in y and z, just what is needed for a projection into the yz-plane.

The six iterated triple integrals can now be given as

- v1 := Tripleint(1, x = y..sqrt(1 - z^2), y = 0..sqrt(1 - z^2), z = 0..1);
  v2 := Tripleint(1, x = y..sqrt(1 - y^2), z = 0..sqrt(1 - y^2), y = 0..1);
  v3 := Tripleint(1, y = 0..x, x = 0..sqrt(1 - z^2), z = 0..1);
  v4 := Tripleint(1, y = 0..x, z = 0..sqrt(1 - x^2), x = 0..1);
  v5 := Tripleint(1, z = 0..sqrt(1 - x^2), x = y..1, y = 0..1);
  v6 := Tripleint(1, z = 0..sqrt(1 - x^2), y = 0..x, x = 0..1);

$$vl := \int_0^1 \int_0^{\sqrt{1-z^2}} \int_y^{\sqrt{1-z^2}} 1 \, dx \, dy \, dz$$

$$v2 := \int_0^1 \int_0^{\sqrt{1-y^2}} \int_y^{\sqrt{1-y^2}} 1 \, dx \, dz \, dy$$

$$v3 := \int_0^1 \int_0^{\sqrt{1-z^2}} \int_0^x 1 \, dy \, dx \, dz$$

$$v4 := \int_0^1 \int_0^{\sqrt{1-x^2}} \int_0^x 1 \, dy \, dz \, dx$$

$$v5 := \int_0^1 \int_y^1 \int_0^{\sqrt{1-x^2}} 1 \, dz \, dx \, dy$$

$$v6 := \int_0^1 \int_0^x \int_0^{\sqrt{1-x^2}} 1 \, dz \, dy \, dx$$

- `value([v.(1..6)]);`

$$\left[\frac{1}{3}, \frac{1}{3}, \frac{1}{3}, \frac{1}{3}, \frac{1}{3}, \frac{1}{3} \right]$$

Example 7

Find the volume inside the upper hemisphere of radius 16 but below the cone $z = \sqrt{(x^2 + y^2)}$.

Spherical coordinates are most appropriate here. Since physics and math texts differ on the definition of these coordinates, we state clearly our notation. The transformation equations are

- `x  =  rho*cos(theta)*sin(phi);`
  `y  =   rho*sin(theta)*sin(phi);`
  `z  = rho*cos(phi);`

$$x = \rho \, \cos(\theta) \, \sin(\phi)$$

$$y = \rho \, \sin(\theta) \, \sin(\phi)$$

$$z = \rho \, \cos(\phi)$$

The reader is urged to resist the temptation to use "fi" for the angle "phi." In Maple, **fi** is a reserved word (it ends **if** statements), and its use as a variable has unpleasant side effects. The angle theta measures rotation around the z-axis and lies in the interval $[0, 2\pi]$, while the angle phi is measured from the z-axis to the radius vector and lies in the interval $[0, \pi]$. The element of volume in spherical coordinates is

$$\rho^2 \sin(\phi) \, d\rho \, d\theta \, d\phi$$

In spherical coordinates the upper hemisphere is given by rho = 4, while the cone is given by phi = $\pi/4$. The iterated triple integral giving the volume is

• v := Tripleint(rho^2*sin(phi), rho = 0..4, phi = Pi/4..Pi/2, theta = 0..2*Pi);

$$v := \int_0^{2\pi} \int_{\frac{1}{4}\pi}^{\frac{1}{2}\pi} \int_0^4 \rho^2 \sin(\phi) \, d\rho \, d\phi \, d\theta$$

• V := value(v);

$$V := \frac{64}{3} \sqrt{2} \, \pi$$

Example 8

Find the z-coordinate of the centroid of the region of Example 7.

The required z-coordinate is given by $[\int_R z \, dv]/V$, where V is the volume computed in Example 7 and R is the region between the cone and the hemisphere.

• M:=Tripleint(rho*cos(phi)*rho^2*sin(phi),rho=0..4,theta=0..2*Pi,phi=Pi/4..Pi/2);

$$M := \int_{\frac{1}{4}\pi}^{\frac{1}{2}\pi} \int_0^{2\pi} \int_0^4 \rho^3 \cos(\phi) \sin(\phi) \, d\rho \, d\theta \, d\phi$$

• value(M)/V;

$$\frac{3}{4} \sqrt{2}$$

Exercises

1. By a Riemann sum, find the approximate value of the double integral $\iint\limits_{R} xy \, dA$ over the

    region $R = \{(x, y) \mid 0 \le x \le 2, \ 0 \le y \le 2\}$.   Partition with n subintervals along each axis.

    (a) Use **leftsum** with $n = 5$.   **(c)** Use **leftsum** with $n = 20$.

    (b) Use **rightsum** with $n = 5$.   **(d)** Use **rightsum** with $n = 20$.

    (e) Use **sum** with $n = 10$. (Evaluate the function at the lower righthand corner of each
       rectangle.)

    (f) Use **leftsum**, taking the limit as $n \to \infty$.

    (g) Use **rightsum**, taking the limit as $n \to \infty$.

    (h) Evaluate $\iint\limits_{R} xy \, dA$ as an iterated integral using the command **int**, and compare
       with the results in parts **(f)** and **(g)**.

    (i) Evaluate the double integral using the commands **Doubleint** and **value**.

2. Repeat Exercise 1 for the double integral $\iint\limits_{R} (5 - x - 2y) \, dA$, where R is the region defined

    by $R = \{(x, y) \mid 0 \le x \le 2, \ -1 \le y \le 1\}$.

3.   Use the command **matrixplot** to show an approximation to the volume of the solid
    given by the double integral of Exercise 2. Your picture should contain 25 "blocks"
    whose altitudes are determined by the value of 5 - x - 2y at the
    "left-hand corner" of each base.

4. Plot the solid that is bounded above by the surface $f(x,y) = \sqrt{x^3 + y^3}$ and below by $R = \{(x, y, 0) \mid 1 \le x \le 2, \ -1 \le y \le 1\}$. Estimate the volume of this solid by computing an
    appropriate Riemann sum and by using Maple's built-in numeric integrator.

5.   Use the command **matrixplot** to show an approximation to the volume of the solid
    given by the double integral of Exercise 4. Your picture should contain 16 "blocks"
    whose altitudes are determined by $\sqrt{x^3 + y^3}$ at the "right-hand corner" of each base.

6. Repeat Exercise 4 for $f(x,y) = \sqrt{36 - 4x^2 - y^2}$ and $R = \{(x, y) \mid 0 \leq x \leq 2, \ 1 \leq y \leq 2\}$.

7. (a) Plot the region over which the integration takes place in $\displaystyle\int_0^1 \int_0^{2x} x\, y \, dy \, dx$.

    (b) Estimate the value of this iterated integral by computing an appropriate Riemann sum.

    (c) Find the exact value of the iterated integral.

    (d) Reverse the order of integration and again evaluate the resulting iterated integral.

8. Repeat Exercise 5 for $\displaystyle\int_0^{1.5} \int_{x^2}^{2x} x\, y^2 \, dy \, dx$.

9. Find the area of the region in the first quadrant that lies inside the curve which is given in polar coordinates by $r = 4\sqrt{2 - \cos 2\theta}$.

10. Find the volume of the solid that is bounded above by the surface whose representation in cylindrical coordinates is $z = 4 - \sin\theta \, \cos\theta$ and below by the region enclosed by $\{(r, \theta, 0) \mid r = 1 + \cos 2\theta\}$.

11. Find or estimate the center of gravity of the lamina

    (a) with density $\delta(x,y) = \sqrt{x^2 + y^3}$, and which is bounded by $x = 0$, $x = 2$, $y = 0$, and $y = 3$.

    (b) with density $\delta(x,y) = \sqrt{xy + \sin^2 x}$, and which is bounded by $y = \sin x$ and $y = 4x - x^2$.

12. Find or estimate the surface area of

    (a) the portion of the plane $z = 12 - 3x - 4y$ in the first octant.
    (b) the portion of $z = \sqrt{x^2 + y^2}$ above the region in the xy-plane that is bounded by $y = x^2$ and $y = 2x$.

    (c) the portion of $z = \sqrt{x^2 + y^3}$ above the region in the xy-plane that is bounded by $y = x^2$, $y = 0$, and $x = 3/2$.

(d) the portion of $z = \cos y \, e^{-\frac{\sqrt{x^2 + y^2}}{10}}$ above the region in the xy-plane that is bounded by $x = 0$, $x = 1$, $y = 0$, and $y = \frac{\pi}{2}$.

13. Find or estimate the centroid of the region in the first quadrant

(a) under $y = e^{-x}$.   (b) under $y = e^{-x^2}$.

14. Use triple integrals in the rectangular, cylindrical, or spherical coordinate system to find or estimate the volume of the solid described.

(a) The solid in the first octant which is bounded above by $z = y$, below by $z = 0$, and on the sides by $x = y^3$ and $x = 8$.

(b) The solid in the first octant which is bounded by the cylinder $x^2 + y^2 = 4$ and the plane $z = \frac{1}{2} y$.

(c) The solid in the first octant which is bounded by the cylinder $x^2 + y^2 = 4x$ and the surface $z = 5 - \sqrt{x^2 + y^2}$.

(d) The solid in the first octant which is inside the sphere $x^2 + y^2 + z^2 = 25$ and outside the surface $z = \sqrt{x^2 + y^2}$.

(e) The solid in the first octant which is inside $(x - 1)^2 + (y - 2)^2 + z^2 = 25$ and inside the surface $z = \sqrt{x^2 + y^2 + 1}$.

Projects

1. Enter each of the following commands. Compare the answers and explain the differences or similarities.

(a) int(c * x^2, x);   (b) int (y * x^2, x);

2. A solid has its top given by $z = 5 - \frac{1}{2}x^2 - \frac{1}{4}y^2$, its bottom by the upper half of $\rho = 2$, and its sides are parallel to the z-axis.

(a) Display the solid with a portion of the side cut away to show part of the base.

(b) Find or estimate the volume of the solid.

3.   A solid has its top given in cylindrical coordinates by $z = 2 + \sin(4\pi r)$, its base is the region inside $r = 1 - \cos\theta$, $z = 0$, and its sides are parallel to the z-axis.

(a) Display the solid with a portion of the side cut away to show part of the base.

(b) Find or estimate the volume of the solid.

4.   (a) Display the surface $f(x,y) = 25 - x^2 - y^2$ and the tangent plane to $f(x,y)$ at the point where $(x, y) = (1, 3)$.

(b) Write an animation that lowers the plane until it contains the point $(1, 3, 10)$.

(c) How far must the original tangent plane be lowered if the solid formed by $f(x,y)$ and the lowered plane is to have volume 1.44?

5.   (a) Write Maple instructions that generate a picture of parallelepipeds that approximate the volume under a surface $f(x,y)$. Your code should do in three dimensions what the command **leftbox** does in two dimensions.

(b) Execute your code for the function $f(x,y) = 4 - \frac{1}{2}x^2 - \frac{1}{4}y^2$, $0 \le x \le 2$, $0 \le y \le 2$.

(c) Write Maple instructions that compute the sum of the volumes of the parallelepipeds generated by your instructions in part (a). Execute your algorithm for the function and domain of part (b).

6.   Apply your instructions from Exercise 5 to the following. Use Maple's **Tripleint** command to set up the appropriate iterated integral, and evaluate by either **value** or **evalf**, as appropriate.

(a) The first octant solid inside $x^2 + y^2 = 16$ and below $z = \sqrt{x^2 + y^2 - 4}$.

(b) The first octant solid satisfying $0 \le x \le 1$, $0 \le y \le \frac{\pi}{2}$, $z \le x^2 \sin y$.

7.   Define $z_1 = 25 - x^2 - y^2$ and $z_2 = 20 - 2x - 6y$.

(a) Display the solid that is enclosed by these two surfaces.

(b) Find the volume of this solid.

8.   An airplane hangar has a rectangular base that is 200 meters wide and 400 meters long. The cross sections, which run the width of the hangar, are semicircles of radius 100 meters.

(a) Generate a picture of the hangar, including the floor. Leave it open at both ends.

(b) Use a double integral to find the volume of the hangar.

(c) The exterior of the hangar is to be painted with a sealing, insulating material that costs $17.83 per square meter. Find the cost of the material required.

9. A second hangar has a rectangular base that is 200 meters wide and 400 meters long. The cross sections, which run the width of the hangar, are not semicircles, but each is three-eighths of a circle. Repeat the questions in Exercise 8.

10. Define $f(x,y) = 72 - \sqrt{x^3 + y^2}$.

(a) Estimate $\displaystyle\int_{-1}^{1}\int_{0}^{3} f(x,y)\,dy\,dx$ by finding a multivariate Taylor expansion $T(x,y)$ of $f(x,y)$ about the origin, and then integrating $T(x,y)$. (see ?mtaylor.)

(b) Display $f(x,y)$ and $T(x,y)$ in the same window for $-1 \leq x \leq 1, \ 0 \leq y \leq 3$. Does $T(x,y)$ appear to give a "good" fit? A measure of the "goodness of fit" could be the value of $\frac{1}{n}\Sigma\Sigma \mid f(x,y) - T(x,y)\mid$, where the summation is taken over n appropriately chosen points.

(c) Use another method to estimate the value of the double integral in part **(a)**, and compare your answers.

11. Work out the details of the following scheme of approximate integration for evaluating double integrals. Then test your method on $\displaystyle\iint_{R} f(x,y)\,dA$, where $f(x,y) = x^2 y$, and $0 \leq x \leq 1, 0 \leq y \leq 2$.

The idea of the method is to replace the surface $z = f(x,y)$ with planar approximations. The grid formed by partitioning the xy-plane with $x_j = x_0 + j\,\Delta x$ and $y_k = y_0 + k\,\Delta y$ contains rectangles with corners at the four points $A = (x_n, y_m)$, $B = (x_{n+1}, y_m)$, $C = (x_{n+1}, y_{m+1})$, $D = (x_n, y_{m+1})$. In general, a plane can be passed through three arbitrary points, not four. Thus, divide the rectangle ABCD into two triangles by the diagonal AC. Pass a plane through the three points that lie in the surface $z = f(x,y)$ above the vertices of the triangle ABC, and pass a plane through the three points in the surface but above the triangle ACD. Find the values of the double integral of $f(x,y)$ over each of these triangles as an approximation of the value of the double integral over the rectangle ABCD. This step should result in a formula analogous to that for Simpson's Rule over one double panel.

Next, devise a formula for the cumulative effect of using the one-rectangle formula over all the rectangles of the grid. This derivation is simplest if it is implemented for a region R that is itself a large rectangle. The difficulties of implementing this strategy for a region with curved boundaries in which the rectangular grid points need not fall on the boundaries should alert you to the inherent challenges of the numeric integration of iterated integrals.

12. The new Super Float golf ball is to be made with a hollow center of radius 1.3 cm. The radius of the golf ball is 2.1 cm, and the nonuniform density of the solid section is cr^2 g/cm^3, where r is the distance from the center of the ball. Find the maximum value of c such that the ball will float in water of density 1 g/cm^3.

13. Jan has a large chocolate-covered doughnut, which is a perfect torus with an outer diameter of 6 inches and an inner (hole) diameter of 2 inches. Unfortunately for Jan, she has to share the doughnut with her younger brother and sister. Laying the doughnut flat on the table, Jan makes two parallel vertical cuts tangent to, and on opposite sides of the hole in the center of the doughnut. That is, she divides the doughnut into three 2-inch wide "slices." She gives one outside piece to her sister and the other outside piece to her brother and keeps the central "slice" (actually two pieces) for herself. What fraction (to three significant figures) of the doughnut and what fraction of the chocolate coating does she keep for herself?

-Technology Review (The Bent)

Chapter 18

Vector Calculus

New Maple Commands for Chapter 18

simplify@simplify	composition; applies simplify twice in succession
simplify(expr,{equation});	simplify expr, using the information in equation
union	set union (union of two sets)
with(linalg);	loads linear algebra package
curl(v,[x,y,z]);	curl of vector **v**, using coordinates [x,y,z]
diverge(v,[x,y,z]);	divergence of vector **v**, using coordinates [x,y,z]

Introduction

The epitome of multivariate calculus might well be the unit on line and surface integrals, and on Green's, Stokes', and the Divergence theorems. The calculations are sophisticated and the nuances of thought arresting. But by far the greatest challenge is visualizing operations and objects in three dimensions. The vector operators of gradient, divergence, and curl are of paramount importance in the applications, and without a clear understanding of the material of Chapter 18 little (classical) applied mathematics is within reach.

Examples

Example 1

Let C be the triangle with vertices (1, 0), (1, 1), and (0, 0). If $P\,dx + Q\,dy = y^2\,dx + x^2\,dy$, obtain the line integral $\int_C P\,dx + Q\,dy$ if C is traversed counterclockwise.

Each segment of the triangle requires its own parametrization. On the segment from (0, 0) to (1, 0) we have y = 0 and x = x. Hence, $\int P\,dx + Q\,dy =$

$$\int_0^1 0\,dx + \int_0^1 x^2\,0$$

which is zero.

On the segment from (1, 0) to (1, 1) we have y = y and x = 1, so that $\int P\,dx + Q\,dy =$

$$\int_0^1 y^2\,0 + \int_0^1 1\,dy$$

which is 1.

On the segment from (1, 1) to (0, 0) we have y = x, so that $\int P\,dx + Q\,dy =$

$$\int_1^0 x^2\,dx + \int_1^0 x^2\,dx = 2\int_1^0 x^2\,dx$$

and this is -2/3. Hence, the desired line integral is 0 + 1 - 2/3 = 1/3.

Example 2

Find the work done by the force field **F** as it moves a particle of unit mass along an ellipse r(t) if **F** = (3 x - 4 y) **i** + (4 x + 2 y) **j** and **r**(t) = 4 cos(t) **i** + 3 sin(t) **j**.

Work done by the force field **F** on the particle is given by the line integral $\int$ **F•dr** where **dr** = dx **i** + dy **j**. Since dx = (dx/dt) dt, and dy = (dy/dt) dt, we can compute the work done as follows.

- with(linalg):
  F := vector([3*x-4*y, 4*x+2*y]);
  r := vector([4*cos(t), 3*sin(t)]);
  Ft := subs(x = r[1],y = r[2], op(F));
  dr := map(diff, r, t);
  Ftdr := dotprod(Ft,dr);

$$F := [\,3\,x - 4\,y \quad 4\,x + 2\,y\,]$$

$$r := [\,4\cos(t) \quad 3\sin(t)\,]$$

$$Ft := [\,12\cos(t) - 12\sin(t) \quad 16\cos(t) + 6\sin(t)\,]$$

$$dr := [\,\text{-}4\sin(t) \quad 3\cos(t)\,]$$

$$Ftdr := \text{-}4\,(12\cos(t) - 12\sin(t))\,\sin(t) + 3\,(16\cos(t) + 6\sin(t))\,\cos(t)$$

The required line integral representing work is then

- int(Ftdr, t = 0..2*Pi);

$$96\,\pi$$

Example 3

Compute the line integral in Example 2 by Green's Theorem in the plane.

Since $\int_C P\,dx + Q\,dy = \iint_R [Q_x - P_y]\,dx\,dy$, with R being the interior of the ellipse C = **r**(t), we can interpret Q_x - P_y as the k-component of the curl of the vector $\mathbf{F} = P\,\mathbf{i} + Q\,\mathbf{j} + 0\,\mathbf{k}$.

- `QxPy  :=  curl(vector([F[1],F[2],0]),[x,y,z]);`

$$QxPy := [0 \quad 0 \quad 8]$$

Thus, $\iint R$ [Qx - Py] dx dy = 8 times the area of the ellipse. But the area of the ellipse is πab, where a = 4 and b = 3, so again, the line integral is again found to be 96π.

Example 4

Compute the surface integral of f(x,y,z) = x y + x z + y z on the surface S, the top half of the cylinder $y^2 + z^2 = 4$, for which $0 \le x \le 5$ and $z \ge 0$.

Since z = $\sqrt{(4 - y^2)}$ on the top half of the cylinder, the element of surface area dA is given by $\sqrt{(1 + (z_x)^2 + (z_y)^2)}$ dx dy. Thus,

- `f := x*y + x*z + y*z;`
  `q := sqrt(4 - y^2);`
  `q1 := sqrt(1 + diff(q,x)^2 + diff(q,y)^2);`

$$f := x\,y + x\,z + y\,z$$

$$q := \sqrt{4 - y^2}$$

$$q1 := \sqrt{1 + \frac{y^2}{4 - y^2}}$$

- `q2 := simplify(q1, {q^2 = z^2});`

$$q2 := 2\sqrt{\frac{1}{z^2}}$$

The simplification was done in such a way as to take into account the equation of the surface, namely, $y^2 + z^2 = 4$. So the element of surface area is dA = (2/z) dx dy. That means the integrand is

- `q3 := expand(f*2/z);`

$$q3 := 2\,\frac{x\,y}{z} + 2\,x + 2\,y$$

But the integration is over the xy-plane so we need to replace z with z(x,y).

• q4 := subs(z = q, q3);

$$q4 := 2\,\frac{x\,y}{\sqrt{4-y^2}}+2\,x+2\,y$$

The integrand, q4, is now in a suitable form for the required integrations to be carried out.

• with(student):

• q5 := Doubleint(q4, x = 0..5, y = -2..2);

$$q5 := \int_{-2}^{2}\int_{0}^{5} 2\,\frac{x\,y}{\sqrt{4-y^2}}+2\,x+2\,y\,dx\,dy$$

• value(q5);

$$100$$

Example 5

Verify Stokes' Theorem for the right hemisphere $y = \sqrt{(9 - x^2 - z^2)}$ and the vector field **F** given by

• F := vector([3*y^2 - 2*x*z, x^2*z + 5*y, 3*y*z]);

$$F := [3\,y^2 - 2\,x\,z \quad x^2 z + 5\,y \quad 3\,y\,z]$$

The element of surface area dA is given by $\sqrt{(1 + (y_x)^2 + (y_z)^2)}\ dx\ dz$. Hence, we compute

• Y := sqrt(9 - x^2 - z^2);
  q6 := sqrt(1 + diff(Y,x)^2 + diff(Y,z)^2);

$$Y := \sqrt{9 - x^2 - z^2}$$

$$q6 := \sqrt{1 + \frac{x^2}{9 - x^2 - z^2} + \frac{z^2}{9 - x^2 - z^2}}$$

• q7 := simplify(q6, {Y^2 = y^2});

$$q7 := 3\sqrt{\frac{1}{y^2}}$$

Clearly, dA = (3/y) dx dz.   Next, we need to obtain a unit normal pointing outward on the surface. Since the surface is a hemishpere, a normalized gradient vector should suffice.

- N := grad(x^2 + y^2 + z^2 - 9, [x,y,z]);

$$N := [\,2\,x \quad 2\,y \quad 2\,z\,]$$

- NN := evalm(N/sqrt(dotprod(N,N)));

$$NN := \left[\frac{x}{\sqrt{x^2+y^2+z^2}} \quad \frac{y}{\sqrt{x^2+y^2+z^2}} \quad \frac{z}{\sqrt{x^2+y^2+z^2}}\right]$$

- NNN := map(simplify@simplify, NN, {x^2 + y^2 + z^2 = 9});

$$NNN := \left[\frac{1}{3}x \quad \frac{1}{3}y \quad \frac{1}{3}z\right]$$

Simplifying NN with respect to the condition that the normal be on the surface of the hemisphere is a rare instance where **simplify** has to be applied twice to get the desired simplification. This repetition of the **simplify** command is done by the composition operator @. We now can obtain the curl of **F** dotted with the unit normal (NNN in Maple).

- q8 := dotprod(curl(F, [x,y,z]), NNN);

$$q8 := \frac{1}{3}(3\,z - x^2)\,x - \frac{2}{3}xy + \frac{1}{3}(2\,xz - 6\,y)\,z$$

The integrand for the surface integral in Stokes' Theorem is therefore

- q9 := expand(q8 * (3/y));

$$q9 := 3\frac{xz}{y} - \frac{x^3}{y} - 2\,x + 2\frac{xz^2}{y} - 6\,z$$

The integration takes place in the xz-plane over the circle x² + z² = 9, so we must replace y with √(9 - x² - z²) in q9. And the only sensible way to carry out this integration is to impose polar coordinates on this circle by setting x = r cos(t) and z = r sin(t). This makes y become √(9 - r²), the integrand become

- q10 := subs(x = r*cos(t), z = r*sin(t), y = sqrt(9 - r^2), q9)*r;

$$q10 := \left(3 \frac{r^2 \cos(t) \sin(t)}{\sqrt{9 - r^2}} - \frac{r^3 \cos(t)^3}{\sqrt{9 - r^2}} - 2 r \cos(t) + 2 \frac{r^3 \cos(t) \sin(t)^2}{\sqrt{9 - r^2}} - 6 r \sin(t) \right) r$$

and the surface integral become

- q11 := Doubleint(q10, r = 0..3, t = 0..2*Pi);

$$q11 := \int_0^{2\pi} \int_0^3 \left(3 \frac{r^2 \cos(t) \sin(t)}{\sqrt{9 - r^2}} - \frac{r^3 \cos(t)^3}{\sqrt{9 - r^2}} - 2 r \cos(t) + 2 \frac{r^3 \cos(t) \sin(t)^2}{\sqrt{9 - r^2}} - 6 r \sin(t) \right) r \, dr \, dt$$

- value(q11);

$$0$$

The shrewd reader would have anticipated this outcome by observing that each term in the integrand contained an odd power of either sin(t) or cos(t). This suggests we will also obtain zero for the line integral around the bounding circle, the circle of radius 3 in the xz-plane. We need a radius vector **r** to describe this circle. From this vector we obtain **dr**.

- r := vector([3*cos(t), 0, 3*sin(t)]);

$$r := [3 \cos(t) \quad 0 \quad 3 \sin(t)]$$

- dr := map(diff, r, t);

$$dr := [-3 \sin(t) \quad 0 \quad 3 \cos(t)]$$

The integrand of the line integral in Stokes' Theorem is **F•dr**, but we need to express this result in terms of t.

- q12 := subs(x = r[1], y = r[2], z = r[3], dotprod(F,dr));

$$q12 := 54 \cos(t) \sin(t)^2$$

Finally, then, the line integral.

- int(q12, t = 0..2*Pi);

$$0$$

Example 6

Verify the Divergence Theorem for the field **F** given below and for the region between the surfaces $z = 2 - \sqrt{(x^2 + y^2)}$ and $z = x^2 + y^2$.

- F := vector([2*x^2, 3*y^2, 4*z]);

$$F := [2\,x^2 \quad 3\,y^2 \quad 4\,z]$$

The prevalence of $x^2 + y^2$ suggests cylindrical coordinates, so that the surfaces are now given by $z = 2 - r$ and $z = r^2$. Eliminating r from these two equations informs us of the height of the intersection of these two surfaces.

- solve({z = 2 - r, z = r^2}, {z, r});

$$\{r = 1, z = 1\}, \{z = 4, r = -2\}$$

The intersection takes place at height $z = 1$ along a circle of radius $r = 1$. We draw a picture of this region.

- with(plots):

- p1 := cylinderplot(sqrt(z), t = 0..2*Pi, z = 0..1):
  p2 := cylinderplot(2-z, t = 0..2*Pi, z = 1..2):
  display([p1, p2]);

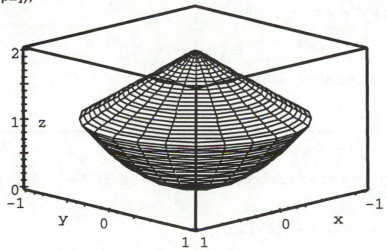

The volume integral in the Divergence Theorem requires the divergence of the field **F**. The symmetry suggests setting up the integral in cylindrical coordinates. Thus,

- q13 := {x = r*cos(t), y = r*sin(t)};

$$q13 := \{x = r\cos(t), y = r\sin(t)\}$$

- q14 := diverge(F, [x, y, z]);

$$q14 := 4\,x + 6\,y + 4$$

- q15 := subs(q13, q14);

$$q15 := 4\,r\cos(t) + 6\,r\sin(t) + 4$$

- q16 := Tripleint(q15*r, z = r^2..2 - r, r = 0..1, t = 0..2*Pi);

$$q16 := \int_0^{2\pi} \int_0^1 \int_{r2}^{2-r} (4\,r\cos(t) + 6\,r\sin(t) + 4)\,r\,dz\,dr\,dt$$

- value(q16);

$$\frac{10}{3}\pi$$

The surface integral in the Divergence Theorem must be computed over each of the two surfaces separately. For each surface we will need an outward unit normal and an appropriate expression for the element of surface area, dA. In each case we will use the gradient to generate a normal vector. A gradient vector for the bottom surface is

- N1 := grad(z - x^2 - y^2, [x, y, z]);

$$N1 := [\text{-}2\,x \quad \text{-}2\,y \quad 1]$$

- N2 := evalm(-N1/sqrt(dotprod(N1,N1)));

$$N2 := \left[2\frac{x}{\sqrt{4\,x^2 + 4\,y^2 + 1}} \quad 2\frac{y}{\sqrt{4\,x^2 + 4\,y^2 + 1}} \quad -\frac{1}{\sqrt{4\,x^2 + 4\,y^2 + 1}} \right]$$

The element of surface area for this bottom surface is $\sqrt{(1 + (zx)2 + (zy)2)}$ dx dy, so that we are moved to compute

- q17 := sqrt(1 + (2*x)^2 + (2*y)^2);

$$q17 := \sqrt{4\,x^2 + 4\,y^2 + 1}$$

The surface integral requires **F•N2** dA, but Maple's **dotprod** command produces unacceptable results in light of the radicals present. Thus, we perform the dot product by summing products of the components of **F** and **N2**,

• q18 := expand(sum(F[k]*N2[k], k = 1..3)*q17);

$$q18 := 4\,x^3 + 6\,y^3 - 4\,z$$

This integration takes place over the unit circle in the xy-plane, so again, polar coordinates are appropriate.

• q19 := subs(q13 union {z = r^2}, q18);

$$q19 := 4\,r^3 \cos(t)^3 + 6\,r^3 \sin(t)^3 - 4\,r^2$$

• q20 := Doubleint(q19*r, r = 0..1, t = 0..2*Pi);

$$q20 := \int_0^{2\pi} \int_0^1 (4\,r^3 \cos(t)^3 + 6\,r^3 \sin(t)^3 - 4\,r^2)\,r\,dr\,dt$$

• q21 := value(q20);

$$q21 := -2\,\pi$$

Repeating these calculations on the top surface leads to

• M1 := grad(z + sqrt(x^2 + y^2) -2, [x, y, z]);

$$M1 := \left[\frac{x}{\sqrt{x^2+y^2}}\quad \frac{y}{\sqrt{x^2+y^2}}\quad 1 \right]$$

• m := normal(sum(M1[k]^2, k = 1..3));

$$m := 2$$

• M2 := evalm(M1/sqrt(m));

$$M2 := \left[\frac{1}{2}\frac{\sqrt{2}\,x}{\sqrt{x^2+y^2}}\quad \frac{1}{2}\frac{\sqrt{2}\,y}{\sqrt{x^2+y^2}}\quad \frac{1}{2}\sqrt{2} \right]$$

• q22 := 2 - sqrt(x^2 + y^2);

$$q22 := 2 - \sqrt{x^2+y^2}$$

• q23 := sqrt(1 + diff(q22,x)^2 + diff(q22,y)^2);

$$q23 := \sqrt{\frac{x^2}{x^2+y^2}+\frac{y^2}{x^2+y^2}+1}$$

• q24 := normal(q23);

$$q24 := \sqrt{2}$$

Thus, the element of surface area on the top surface is $\sqrt{2}$ dx dy. The complete integrand for the surface integral over the top surface is **F•M2** dA.

• q25 := expand(sum(F[k]*M2[k], k = 1..3)*q24);

$$q25 := 2\frac{x^3}{\sqrt{x^2+y^2}}+3\frac{y^3}{\sqrt{x^2+y^2}}+4z$$

Again, this integral is over the unit circle in the xy-plane, so we express z as z(x,y) and switch to polar coordinates.

• q26 := subs(z = q22, q25);

$$q26 := 2\frac{x^3}{\sqrt{x^2+y^2}}+3\frac{y^3}{\sqrt{x^2+y^2}}+8-4\sqrt{x^2+y^2}$$

• q27 := simplify(subs(q13, q26), symbolic);

$$q27 := 2\,r^2\cos(t)^3+3\,r^2\sin(t)-3\,r^2\sin(t)\cos(t)^2+8-4\,r$$

• q28 := Doubleint(q27*r, r = 0..1, t = 0..2*Pi);

$$q28 := \int_0^{2\pi}\int_0^1 \left(2\,r^2\cos(t)^3+3\,r^2\sin(t)-3\,r^2\sin(t)\cos(t)^2+8-4\,r\right)r\,dr\,dt$$

- q29 := value(q28);

$$q29 := \frac{16}{3}\pi$$

The surface integrals are then the sum of q21 and q29

- q21 + q29;

$$\frac{10}{3}\pi$$

And this is the same value we computed for the volume integral earlier.

Exercises

The following paths are used in Exercises 1 and 2:

(A) C: The path starts at the point (0, 0), moves straight to (1, 0), then follows the arc of the circle $x^2 + y^2 = 1$ in a counterclockwise direction to the point (0, 1).

(B) C: The path starts at the point (0, 0), moves straight to (1, 1), then follows the arc of the circle $x^2 + y^2 = 2$ in a clockwise direction to the point ($\sqrt{2}$, 0).

(C) C: The triangular path that starts at the point (1, 0), moves to (0, 1), then to (-1, 0), and then to (1, 0).

(D) C: The path starts at the point (0, 0) and follows $y = x^3$ to the point (2, 8).

(E) C: The path is the boundary of an upper semi-disk: Start at the point (1, 0), move on the circular arc of radius 1 through (0, 1) to (-1, 0), and then straight to (1, 0).

(F) C: The path starts at the point (0, 0), moves straight to (0, 1), and then follows $y = e^x$ to (1, e).

(G) C: The path starts at the point (3, 0) and moves in a clockwise direction around the circle of radius 1 having center at the point (2, 0).

1. Find or approximate the value of the line integral $\int_C x\, y\, dx + x^2\, dy$ with C defined above:

   (a) in A   **(b)** in B   **(c)** in C
   (d) in D   **(e)** in E   **(f)** in F
   (g) in G.

2. Find (or approximate) the value of the line integral $\int_C e^y\, dx + \sin x\, dy$ with C defined above:

   (a) in A   **(b)** in B   **(c)** in C
   (d) in D   **(e)** in E   **(f)** in F
   (g) in G.

In Exercises 3 and 4, find or approximate the value of the surface integral $\iint_\sigma g(x,y,z) \, dS$.

3. $g(x,y,z) = y$ and σ is the portion of the plane $x + y + z = 3$ in the first octant.

4. $g(x,y, z) = z^2 + 4$ and σ is the portion of the cone $z^2 = x^2 + y^2$ with $0 \le z \le 3$.

In Exercises 5 and 6, use the Divergence theorem to evaluate $\iint_\sigma \mathbf{F} \cdot \mathbf{n} \, dS$ where $\mathbf{n}$ is the outer unit normal to σ.

5. $\mathbf{F}(x,y,z) = [x, -y, 5 z]^t$, and σ is the surface bounded by $x = 0$, $x = 2$, $y = 0$, $y = 2$, $z = 0$, and $z = 4$.

6. $\mathbf{F}(x,y,z) = [x^2, y^2, z^2]^t$, and σ is the surface of the solid formed on the side by the cylinder $x^2 + y^2 = 4$, below by $z = 0$, and above by $z = 4$.

In Exercises 7 and 8, use Stokes' Theorem to evaluate $\int_C \mathbf{F} \cdot d\mathbf{r}$.

7. $\mathbf{F}(x,y,z) = [z, x, y]^t$, and C is the circle $x^2 + y^2 = 4$ in the plane $z = 0$.

8. $\mathbf{F}(x,y,z) = [z - x, x - y, y - z]^t$, and C is the triangle having vertices at (1, 0, 0), (0, 2, 0), and (0, 0, 1).

Projects

1.   Plot the vector field

   (a) $\mathbf{F}(x,y) = [x, 3 y]^t$ b) $\mathbf{F}(x,y) = [y, x]^t$.

2.   In the same window, plot the vector field $\mathbf{F}(x,y) = [2x, y]^t$ and the curve C defined by $\mathbf{r}(t) = [\cos t, 2 \sin t]^t$ for $0 \le t \le 2\pi$. Find the work done by the vector field $\mathbf{F}(x,y)$ along the curve C.

3. Evaluate the line integral $\oint_C x\,y\,dx + 2(x^2 + y^2)\,dy$ when C is the ellipse with foci at (0, -4) and (0, 4), and the length of the semi-major axis is 7.

4. The shape of a wire is given by $r(t) = [\cos t, \sin t, t^2]^t$ for $0 \le t \le 2\pi$. If the linear density of the wire is given by $\delta(x,y) = x^2 + 4\,|y|$, find the mass of the wire.

5. Define R to be the region that is inside of $(x - 4)^2 + y^2 = 9$ and outside of $(x - 3)^2 + (y - 1)^2 = 1$. Plot R. Evaluate $\oint_C -y\,dx + 4\,x\,dy$ when C is the boundary of R.

Chapter 19

Second-Order Differential Equations

New Maple Commands for Chapter 19

collect(expr,[A,B]);	in expr, group terms having factors A and B
dsolve(q,y(x));	solve differential equation q for y(x)
dsolve({q,y(0)=A,D(y)(0)=B},y(x));	solve DE q for y(x) and apply initial conditions y(0) = A, y'(0) = B
simplify(expr,{Equations});	simplify expr, using the information in Equations

---

Introduction

What science knows it learned by measuring changes. One can perhaps argue that distance is not known via an increment, but time, motion, and even the dials on meters and instruments are studied through observed changes. The mathematical language of change is the derivative, so it is natural in a fundamental way that the laws of nature are encased in differential equations.

A differential equation is a relationship between derivatives. But derivatives are measures of relative changes, and differential equations are statements about the connections between changes in varying quantities.

Chapter 19 is a glimpse at the way the objects called differential equations are used to represent such commonplace phenomena as a mass dangling at the end of a spring. In particular, wc will examine linear differential equations, those that permit causes to be added together without interference between cause and effect. The literature on differential equations is very extensive and covers nearly 200 years of mathematical discovery and investigation. We reflect but a shadow of this wealth in the examples presented here.

Examples

Example 1

Show that the linear first-order homogeneous differential equation $y'(x) + 2\,y(x) = 0$ has a solution in terms of an exponential.

- q := diff(y(x),x) + 2*y(x) = 0;

$$q := \left(\frac{\partial}{\partial x}y(x)\right) + 2\,y(x) = 0$$

- q1 := dsolve(q, y(x));

321

$$q1 := y(x) = e^{(-2\,x)}\,\_C1$$

This differential equation is separable, and its solution can also be found by separating variables and integrating both sides of the resulting equation. But in either event we will have demonstrated that the solution is given by an exponential term of the form e^{mx}.

Before going further, we issue a caution against naming any of these solutions y. Avoid using as a tag on the left of an assignment (:=) any variable needed as a working variable on the right.

Example 2

Show that the second-order linear homogeneous differential equation y"(x) + y'(x) - 2 y(x) = 0 has its solution given in terms of exponentials.

• q2 := diff(y(x),x,x) + diff(y(x),x) - 2*y(x) = 0;

$$q2 := \left(\frac{\partial^2}{\partial x^2}y(x)\right) + \left(\frac{\partial}{\partial x}y(x)\right) - 2\,y(x) = 0$$

• q3 := dsolve(q2, y(x));

$$q3 := y(x) = \_C1\,e^{x} + \_C2\,e^{(-2\,x)}$$

As promised, the solution is given in terms of exponentials. But suppose we did not have a **dsolve** command to rely on. Where did the solution in q3 come from?

Example 3

Show that the differential equation in Example 2 can be solved by assuming a solution of the form y = e^{mx} and using the differential equation itself to determine the constant m.

• q4 := subs(y(x) = exp(m*x), q2);

$$q4 := \left(\frac{\partial^2}{\partial x^2}e^{(m\,x)}\right) + \left(\frac{\partial}{\partial x}e^{(m\,x)}\right) - 2\,e^{(m\,x)} = 0$$

• q5 := value(q4);

$$q5 := m^2\,e^{(m\,x)} + m\,e^{(m\,x)} - 2\,e^{(m\,x)} = 0$$

• q6 := factor(q5);

$$q6 := e^{(m\,x)}\,(m+2)\,(m-1) = 0$$

If e^{mx} is to satisfy the differential equation q2, the value of m will have to be either 1 or -2. Hence, e^x and e^{-2x} are both solutions of the differential equation. The general solution is then a linear combination of these two building-block solutions. A pair of such distinct (independent) building-block solutions is called a *fundamental set of solutions*.

Example 4

Show that a linear combination of any two solutions of a second-order linear homogeneous differential equations is itself a solution of the equation, an observation known as the Principle of Superposition.

Enter a generic second-order linear homogeneous differential equation.

• q7 := a*diff(y(x),x,x) + b*diff(y(x),x) + c*y(x) = 0;

$$q7 := a\left(\frac{\partial^2}{\partial x^2}\,y(x)\right) + b\left(\frac{\partial}{\partial x}\,y(x)\right) + c\,y(x) = 0$$

Tell Maple that u(x) and v(x) are to be considered solutions of this differential equation.

• q8 := subs(y(x) = u(x), q7);
  q9 := subs(y(x) = v(x), q7);

$$q8 := a\left(\frac{\partial^2}{\partial x^2}\,u(x)\right) + b\left(\frac{\partial}{\partial x}\,u(x)\right) + c\,u(x) = 0$$

$$q9 := a\left(\frac{\partial^2}{\partial x^2}\,v(x)\right) + b\left(\frac{\partial}{\partial x}\,v(x)\right) + c\,v(x) = 0$$

Form the linear combination A u(x) + B v(x) and see if it also satisfies the differential equation.

• q10 := subs(y(x) = A*u(x) + B*v(x), q7);

$$q10 := a\left(\frac{\partial^2}{\partial x^2}\,(A\,u(x) + B\,v(x))\right) + b\left(\frac{\partial}{\partial x}\,(A\,u(x) + B\,v(x))\right) + c\,(A\,u(x) + B\,v(x)) = 0$$

• q11 := value(q10);

$$q11 := a\left(A\left(\frac{\partial^2}{\partial x^2}u(x)\right) + B\left(\frac{\partial^2}{\partial x^2}v(x)\right)\right) + b\left(A\left(\frac{\partial}{\partial x}u(x)\right) + B\left(\frac{\partial}{\partial x}v(x)\right)\right)$$
$$+ c\left(A\,u(x) + B\,v(x)\right) = 0$$

We are looking for evidence that the left-hand side is identically zero whenever equations q8 and q9 are satisfied. If we group terms differently, perhaps it might just jump out at us.

- collect(q11, [A, B]);

$$\left(a\left(\frac{\partial^2}{\partial x^2}u(x)\right) + b\left(\frac{\partial}{\partial x}u(x)\right) + c\,u(x)\right)A + \left(a\left(\frac{\partial^2}{\partial x^2}v(x)\right) + b\left(\frac{\partial}{\partial x}v(x)\right) + c\,v(x)\right)B = 0$$

By grouping against A and B, we see that each constant is multiplied by an expression known to be zero in light of equations q8 and q9. But we can have Maple demonstrate that with finality if we simplify q11 in light of equations q8 and q9.

- simplify(q11, {q8, q9});

$$0 = 0$$

We have therefore demonstrated that the sum of two solutions is itself a solution.

Example 5

Solve the differential equation y"(x) + 6 y'(x) + 13 y(x) = 0.

Begin with a solution based on **dsolve** and then show where that solution came from.

- q12 := diff(y(x),x,x) + 6*diff(y(x),x) + 13*y(x) = 0;

$$q12 := \left(\frac{\partial^2}{\partial x^2}y(x)\right) + 6\left(\frac{\partial}{\partial x}y(x)\right) + 13\,y(x) = 0$$

- q13 := dsolve(q12, y(x));

$$q13 := y(x) = \_C1\,e^{(-3\,x)}\cos(2\,x) + \_C2\,e^{(-3\,x)}\sin(2\,x)$$

Please observe that the distinct building-block solutions are e^{-3x} cos(2x) and e^{-3x} sin(2x). In spite of this we will still look for a solution of the form e^{mx}.

- q14 := subs(y(x) = exp(m*x), q12);

$$q14 := \left(\frac{\partial^2}{\partial x^2} e^{(m\,x)} \right) + 6 \left(\frac{\partial}{\partial x} e^{(m\,x)} \right) + 13\, e^{(m\,x)} = 0$$

- q15 := value(q14);

$$q15 := m^2\, e^{(m\,x)} + 6\, m\, e^{(m\,x)} + 13\, e^{(m\,x)} = 0$$

- q16 := factor(q15);

$$q16 := e^{(m\,x)} (m^2 + 6\, m + 13) = 0$$

The polynomial equation that m must satisfy, called the *characteristic equation*, did not factor. But it is just a quadratic and can be solved in Maple.

- q17 := solve(q16, m);

$$q17 := -3 + 2\,I, -3 - 2\,I$$

The roots, called the *characteristic roots*, are complex numbers. They lead to solutions of the form

- yc1 := exp(q17[1]*x);
  yc2 := exp(q17[2]*x);

$$yc1 := e^{((-3+2\,I)\,x)}$$

$$yc2 := e^{((-3-2\,I)\,x)}$$

The typical calculus student has not encountered imaginary exponentials before. Let's see if Maple assigns them a meaning, and then, if it does, perhaps we can investigate the origins of that meaning.

- q18 := yc1 = evalc(yc1);
  q19 := yc2 = evalc(yc2);

$$q18 := e^{((-3+2\,I)\,x)} = e^{(-3\,x)} \cos(2\,x) + I\, e^{(-3\,x)} \sin(2\,x)$$

$$q19 := e^{((-3-2\,I)\,x)} = e^{(-3\,x)} \cos(2\,x) - I\, e^{(-3\,x)} \sin(2\,x)$$

Equations q18 and q19 express the relations known as Euler's formulas. But recall the Principle of Superposition. The sum (or linear combination) of two solutions is again a solution. Then the linear combinations

- y1 := (yc1 + yc2)/2;
  y2 := (yc1 - yc2)/(2*I);

$$y1 := \frac{1}{2} e^{((-3 + 2 I) x)} + \frac{1}{2} e^{((-3 - 2 I) x)}$$

$$y2 := -\frac{1}{2} I \left(e^{((-3 + 2 I) x)} - e^{((-3 - 2 I) x)} \right)$$

inspired by Euler's formulas, which suggest that these combinations should generate sines and cosines, must themselves be solutions.

- evalc(y1);
  evalc(y2);

$$e^{(-3 x)} \cos(2 x)$$

$$e^{(-3 x)} \sin(2 x)$$

So, if a search for exponential solutions of the form e^{mx} for a linear second-order homogeneous differential equation leads to characteristic roots that are complex, the imaginary exponential solutions generated can be added and subtracted to form new solutions that are "real."

Example 6

Show that $e^{ix} = \cos(x) + i \sin(x)$.

Demonstrations of this fundamenal identity are based on Taylor series representations of each function involved.

- q20 := convert(taylor(exp(I*x), x, 10), polynom);

$$q20 := 1 + I x - \frac{1}{2} x^2 - \frac{1}{6} I x^3 + \frac{1}{24} x^4 + \frac{1}{120} I x^5 - \frac{1}{720} x^6 - \frac{1}{5040} I x^7 + \frac{1}{40320} x^8$$
$$+ \frac{1}{362880} I x^9$$

- q21 := convert(taylor(sin(x), x, 10), polynom);

$$q21 := x - \frac{1}{6} x^3 + \frac{1}{120} x^5 - \frac{1}{5040} x^7 + \frac{1}{362880} x^9$$

- q22 := convert(taylor(cos(x), x, 10), polynom);

$$q22 := 1 - \frac{1}{2}x^2 + \frac{1}{24}x^4 - \frac{1}{720}x^6 + \frac{1}{40320}x^8$$

Using these truncated series to represent the functions e^{ix}, sin(x), and cos(x), check if the identity is valid.

• q23 := q20 - (q22 + I*q21);

$$q23 := I\,x - \frac{1}{6}I\,x^3 + \frac{1}{120}I\,x^5 - \frac{1}{5040}I\,x^7 + \frac{1}{362880}I\,x^9$$
$$- I\left(x - \frac{1}{6}x^3 + \frac{1}{120}x^5 - \frac{1}{5040}x^7 + \frac{1}{362880}x^9\right)$$

• evalc(q23);

$$0$$

A proof that uses infinite series follows along the same lines.

Example 7

Solve the differential equation y"(x) + 4 y'(x) + 4 y(x) = 0.

• q24 := diff(y(x),x,x) + 4*diff(y(x),x) + 4*y(x) = 0;

$$q24 := \left(\frac{\partial^2}{\partial x^2}y(x)\right) + 4\left(\frac{\partial}{\partial x}y(x)\right) + 4\,y(x) = 0$$

• dsolve(q24, y(x));

$$y(x) = e^{(-2\,x)}\_C1 + \_C2\,e^{(-2\,x)}\,x$$

Inspection of this solution shows that e^-2x is an exponential solution, but x e^-2x also appears as a solution. Where did this second solution come from? Should we try y(x) = e^mx?

• q25 := value(subs(y(x) = exp(m*x), q24));

$$q25 := m^2\,e^{(m\,x)} + 4\,m\,e^{(m\,x)} + 4\,e^{(m\,x)} = 0$$

• q26 := factor(q25);

$$q26 := e^{(m\,x)}(m+2)^2 = 0$$

The characteristic roots are -2, and again -2. These roots give only one distinct exponential solution, namely, e^{-2x}. For the other solution, we try something of the form $u(x)\, e^{-2x}$.

• q27 := value(subs(y(x) = u(x)*exp(-2*x), q24));

$$q27 := \left(\frac{\partial^2}{\partial x^2} u(x)\right) e^{(-2\,x)} = 0$$

The equation governing $u(x)$ is just $u''(x) = 0$, and this has the solution $a + bx$; we can therefore form the second solution $(a + bx)e^{-2x}$. But this solution would contain a term $a\, e^{-2x}$ which is not distinct from e^{-2x} already found. So, take $a = 0$ and the simplest second distinct solution is then $x\, e^{-2x}$.

Example 8

Solve the second-order linear nonhomogeneous differential equation $y''(x) + y'(x) - 2\, y(x) = e^{3x}$.

• q28 := diff(y(x),x,x) + diff(y(x),x) - 2*y(x) = exp(3*x);

$$q28 := \left(\frac{\partial^2}{\partial x^2} y(x)\right) + \left(\frac{\partial}{\partial x} y(x)\right) - 2\, y(x) = e^{(3\,x)}$$

• q29 := dsolve(q28, y(x));

$$q29 := y(x) = \frac{1}{10} e^{(3\,x)} + \_C1\, e^x + \_C2\, e^{(-2\,x)}$$

The part of the solution that is just a linear combination of e^x and e^{-2x} is the solution of the homogeneous equation q2 in Example 2. The additional term $e^{3x}/10$ therefore has to be from the nonhomogeneous term on the right side of the differential equation. This term is called a particular solution, whereas the part $\_C1\, e^x + \_C2\, e^{-2x}$ is called the *homogeneous* (complementary in some books) solution. Hence, to solve a nonhomogeneous equation, first solve the homogeneous version to obtain two independent solutions and the ensuing homogeneous solution. Then find a particular solution for the nonhomogeneous equation and sum the particular and homogeneous solutions.

Can the particular solution be found directly? What is the algorithm by which Maple found the particular solution $e^{3x}/10$? We illustrate the method called *Variation of Parameters* in which is sought a particular solution of the form $yp = u(x)\, y1(x) + v(x)\, y2(x)$, where $y1(x)$ and $y2(x)$ are members of a fundamental set of solutions for the homogeneous differential equation, and $u(x)$ and $v(x)$ are unknown multipliers we find by demanding that y_p be a solution of the nonhomogeneous equation. Here, y_p will be

- yp := u(x) * exp(x) + v(x) * exp(-2*x);

$$yp := u(x)\,e^{x} + v(x)\,e^{(-2\,x)}$$

Substitute y_p into the differential equation.

- q30 := eval(subs(y(x) = yp, q28));

$$q30 := \left(\frac{\partial^2}{\partial x^2}u(x)\right)e^{x} + 3\left(\frac{\partial}{\partial x}u(x)\right)e^{x} + \left(\frac{\partial^2}{\partial x^2}v(x)\right)e^{(-2\,x)} - 3\left(\frac{\partial}{\partial x}v(x)\right)e^{(-2\,x)} =$$
$$e^{(3\,x)}$$

We note that we have just one condition with which to determine the two functions u(x) and v(x). And worse, this one condition contains second derivatives of these two unknown functions. We can, and will, impose another condition on equation q30. If it suffices to lead to a solution, then we will rejoice, drawing comfort from the uniqueness theorem which tells us that no matter how a solution is found, if it satisfies the differential equation, then it is the only solution.

To see what additional condition we might profitably impose, examine the derivative of yp.

- diff(yp, x);

$$\left(\frac{\partial}{\partial x}u(x)\right)e^{x} + u(x)\,e^{x} + \left(\frac{\partial}{\partial x}v(x)\right)e^{(-2\,x)} - 2\,v(x)\,e^{(-2\,x)}$$

This first derivative of y_p contains first derivatives of u(x) and v(x). Clearly, taking a second derivative of y_p will lead to second derivatives of u(x) and v(x) as well. If we can somehow prevent even the first derivatives of u(x) and v(x) from surviving, then equation q30 will not have u''(x) or v''(x). Hence, assume u'(x) y_1(x) + v'(x) y_2(x) = 0.

- q31 := diff(u(x),x) * exp(x) + diff(v(x),x) * exp(-2*x) = 0;

$$q31 := \left(\frac{\partial}{\partial x}u(x)\right)e^{x} + \left(\frac{\partial}{\partial x}v(x)\right)e^{(-2\,x)} = 0$$

This means that the derivative of q31 will also be a valid consequence of our new condition.

- q32 := diff(q31, x);

$$q32 := \left(\frac{\partial^2}{\partial x^2} u(x) \right) e^x + \left(\frac{\partial}{\partial x} u(x) \right) e^x + \left(\frac{\partial^2}{\partial x^2} v(x) \right) e^{(-2x)} - 2 \left(\frac{\partial}{\partial x} v(x) \right) e^{(-2x)} = 0$$

The two second derivatives, namely, u"(x) and v"(x), appear in equation q32 in such a way that if we subtract equations q32 and q30 these second derivatives will be eliminated.

• q33 := q30 - q32;

$$q33 := 2 \left(\frac{\partial}{\partial x} u(x) \right) e^x - \left(\frac{\partial}{\partial x} v(x) \right) e^{(-2x)} = e^{(3x)}$$

Equations q31 and q33 contain u'(x) = ∂u/∂x and v'(x) = ∂v/∂x. Hence, we solve these two equations for the derivatives of u(x) and v(x). If we can find these derivatives, we can recover u(x) and v(x) themselves.

• q34 := solve({q31, q33}, {diff(u(x),x), diff(v(x),x)});

$$q34 := \left\{ \frac{\partial}{\partial x} u(x) = \frac{1}{3} \frac{e^{(3x)}}{e^x}, \frac{\partial}{\partial x} v(x) = -\frac{1}{3} \frac{e^{(3x)}}{e^{(-2x)}} \right\}$$

We recover u(x) and v(x) by integration.

• q35 := map(int, q34[1], x);
  q36 := map(int, q34[2], x);

$$q35 := u(x) = \frac{1}{6}(e^x)^2$$

$$q36 := v(x) = -\frac{1}{15}(e^x)^5$$

The particular solution is recovered via the substitution

• simplify(subs(q35, q36, yp));

$$\frac{1}{10} e^{(3x)}$$

Example 9

Find, by the method of *Undetermined Coefficients*, a particular solution for the following nonhomogeneous differential equation.

- q37 := diff(y(x),x,x) + 2*diff(y(x),x) + 10*y(x) = cos(x);

$$q37 := \left(\frac{\partial^2}{\partial x^2} y(x)\right) + 2\left(\frac{\partial}{\partial x} y(x)\right) + 10\, y(x) = \cos(x)$$

Assume a particular solution of the form $y_p = A\cos(x) + B\sin(x)$. Use the differential equation itself to determine the unknown coefficients A and B.

- yp := A*cos(x) + B*sin(x);

$$yp := A\cos(x) + B\sin(x)$$

- q38 := eval(subs(y(x) = yp, q37));

$$q38 := 9\, A\cos(x) + 9\, B\sin(x) - 2\, A\sin(x) + 2\, B\cos(x) = \cos(x)$$

Equation q38 must be an identity in both cos(x) and sin(x). Hence, group the terms multiplying cos(x) and the terms multiplying sin(x) on each side of equation q38. These coefficients of cos(x) and sin(x) on each side of equation q38 must match identically.

- q39 := collect(q38, [cos(x), sin(x)]);

$$q39 := (9\, A + 2\, B)\cos(x) + (9\, B - 2\, A)\sin(x) = \cos(x)$$

Matching coefficients on each side will lead to two equations in A and B.

- q40 := map(coeff, q39, cos(x));
  q41 := map(coeff, q39, sin(x));

$$q40 := 9\, A + 2\, B = 1$$

$$q41 := 9\, B - 2\, A = 0$$

- q42 := solve({q40, q41}, {A, B});

$$q42 := \left\{ B = \frac{2}{85}, A = \frac{9}{85} \right\}$$

We recover the desired particular solution by the following substitution.

- subs(q42, yp);

$$\frac{9}{85}\cos(x) + \frac{2}{85}\sin(x)$$

Finally, we verify that this is the same particular solution Maple would find if we invoked the **dsolve** command.

- q43 := dsolve(q37, y(x));

$$q43 := y(x) = \frac{2}{51}\cos(3\,x)\cos(4\,x) - \frac{1}{102}\cos(3\,x)\sin(4\,x) + \frac{1}{15}\cos(3\,x)\cos(2\,x)$$

$$- \frac{1}{30}\cos(3\,x)\sin(2\,x) + \frac{1}{30}\sin(3\,x)\cos(2\,x) + \frac{1}{15}\sin(3\,x)\sin(2\,x)$$

$$+ \frac{1}{102}\sin(3\,x)\cos(4\,x) + \frac{2}{51}\sin(3\,x)\sin(4\,x) + \_C1\,e^{(-x)}\cos(3\,x)$$

$$+ \_C2\,e^{(-x)}\sin(3\,x)$$

We counteract Maple's propensity for expanding trigonometric functions like sin(2x) to 2 sin(x) cos(x) by using the **combine** command.

- combine(rhs(q43), trig);

$$\frac{9}{85}\cos(x) + \frac{2}{85}\sin(x) + \_C1\,e^{(-x)}\cos(3\,x) + \_C2\,e^{(-x)}\sin(3\,x)$$

In this form of the solution, we can see that the particular solutions agree and that the homogeneous solution is built from the fundamental set {e⁻ˣ cos(3x), e⁻ˣ sin(3x)}.

Example 10

The motion of a damped mass on the end of a spring is described by the initial value problem $y''(t) + y'(t) + 17\,y(t)/4 = 0$, $y(0) = 4$, $y'(0) = -2$. Find and plot the motion y(t), find the maximum displacement of the mass from the equilibrium position, find the time when the mass passes through the equilibrium position for the third time, and plot the solution along with the curves that envelope the solution.

- q44 := diff(y(t),t,t) + diff(y(t),t) + 17*y(t)/4 = 0;

$$q44 := \left(\frac{\partial^2}{\partial t^2}\, y(t)\right) + \left(\frac{\partial}{\partial t}\, y(t)\right) + \frac{17}{4}\, y(t) = 0$$

- q45 := dsolve({q44, y(0) = 4, D(y)(0) = -2}, y(t));

$$q45 := y(t) = 4\, e^{\left(-\frac{1}{2}t\right)} \cos(2\,t)$$

Since we are going to manipulate this solution, we should extract it from q45. Note the use of Y, not y, as a label.

- Y := rhs(q45);

$$Y := 4\, e^{\left(-\frac{1}{2}t\right)} \cos(2\,t)$$

- plot(Y, t = 0..10);

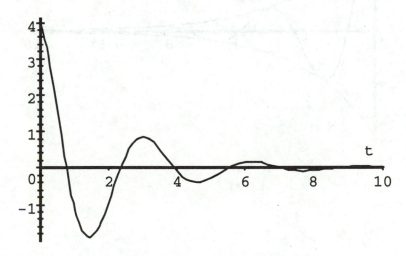

The maximum displacement from equilibrium initially occurs when it is 4. The next maximum occurs at the first turning point on the graph above, which we can find by the usual techniques of optimization.

- q46 := solve(diff(Y, t) = 0, t);

$$q46 := -\frac{1}{2}\arctan\!\left(\frac{1}{4}\right)$$

The time can't be negative. The correct time is $\pi/2$ + q46 = $\pi/2$ - arctan(1/4)/2.

Passages through equilibrium occur when $Y = 0$. Hence,

• q47 := solve(Y = 0, t);

$$q47 := \frac{1}{4}\pi$$

Again, we have the same difficulty with quadrants. The correct answer is $\pi + q47 = 5\pi/4$.

Finally, the envelope of the solution is formed by the curves $\pm 4\, e^{-t/2}$.

• plot({Y, 4*exp(-t/2), -4*exp(-t/2)}, t = 0..10);

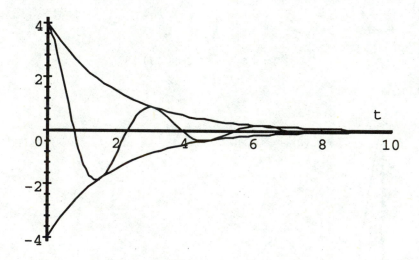

Exercises

In Exercises 1 - 3, find the general solution of the differential equation. Obtain the characteristic equation, the characteristic roots, and a fundamental set of solutions. Then check your results against Maple's **dsolve** command.

1. $y'' - 3y' - 10y = 0$

2. $y'' + 6y' + 9y = 0$

3. $y'' + 4y' + 13y = 0$

In Exercises 4 - 7, solve the differential equations by finding the homogeneous solution and a particular solution. For a right-hand side that is an exponential, try a particular solution that is itself a multiple of that exponential. For a polynomial, use a complete polynomial of the same degree. For one that is a sine or cosine, try a particular solution that contains both a sine and a cosine (so there would be two constants to determine in such a case). Check your results against the output from Maple's **dsolve** command.

4. $y'' - 3y' - 10y = e^x$

5. $y'' - 3y' - 10y = e^{5x}$

6. $y'' - 3y' - 10y = 4\,x^2$

7. $y'' - 3y' - 10y = \cos(x)$

In Exercises 8 - 11, solve the differential equations by obtaining both the homogeneous solution and a particular solution. Obtain the particular solution by Variation of Parameters, and compare your results with those produced by Maple's **dsolve** command.

8. $y'' - y = e^x$

9. $y'' - 4y' + 4y = (x+1)e^x$

10. $y'' + y = x$

11. $y'' + y = \cosh(x)$

In Exercises 12 - 15, the motion of a weight attached to a spring is described by an initial value problem. Obtain the general solution to each using Maple; then determine the two constants in this solution by using the initial conditions. Check this result against the output of Maple's **dsolve** command used to solve the initial value problem.

12. $y'' + 16\,y = 0$, $y(0) = -2$, $y'(0) = 10$. Solve and then plot the motion for $0 \le t \le 4$.

13. $y'' + 50\,y = 0$, $y(0) = 1$, $y'(0) = 10$. Solve and then plot the motion for $0 \le t \le 2$.

14. $y'' + 5\,y' + 6\,y = 0$, $y(0) = 1$, $y'(0) = 2$. Solve and then plot for $0 \le t \le 4$. Find the maximum value of y(t) on the interval $0 \le t \le 4$.

15. $y'' + 4\,y' + 20\,y = 0$, $y(0) = 2$, $y'(0) = 1$. Solve and then plot for $0 \le t \le 3$. Find the greatest distance the weight reaches below the equilibrium point (which is the origin.)

16. A 50 pound weight, attached to the end of a spring, stretches it 6 inches. Find the equation of motion if the system is undamped and the weight is released from a point 4 inches above the equilibrium position. Plot the motion for $0 \le t \le 10$.

17. A 50 pound weight, attached to the end of a spring, stretches it 4 inches. From a point 6 inches above the equilibrium position, the weight is pushed downward with a velocity of 4 feet per second.

    (a) At what time does the weight pass through the equilibrium position heading downward for the first time?

    (b) The second time?

    (c) At what time is the weight 3 inches above the equilibrium position for the first time?

    (d) The second time?

    (e) What is the velocity of the weight when $t = 5$ seconds?

    (f) When is the speed first equal to 2 feet per second in the downward direction?

    (g) In the upward direction?

18. A mass of four slugs is attached to a spring with spring constant $k = 10$ and placed in a medium that has a damping constant $c = 5$. The mass is then released from a point 2 feet below the equilibrium position.

    (a) Find the equation of motion.

    (b) Plot the motion for $0 \le t \le 25$.

    (c) Find the maximum displacement from the equilibrium position for this mass.

19. For an appropriate external force acting on a spring system, the equation of motion can be expressed in the form $y = 4 \sin(t/5)\sin(4t)$. Plot this motion for $0 \le t \le 30$. (Motion of this form is termed *beats*.) Find the maximum displacement of the weight from the equilibrium position. What is the period of this motion? (A very interesting film that shows the tragic results of an external force acting on the Tacoma Narrows Suspension

Bridge is distributed by the National Committee for Fluid Mechanics Films, Educational Services, Inc., Watertown, MA.)

Projects

1. An undamped spring-mass system has mass m = 1 and spring constant k = 2. Initially, the mass is displaced one unit from equilibrium and released. Solve for the motion and use your solution to study the total energy of the system. This total energy is the sum of the kinetic and potential energies, which are, respectively, $\frac{1}{2}mv^2$ and $\frac{1}{2}ky^2$. Find the total energy E and study its behavior as a function of time. (<u>Caution</u>: Maple uses the letter E for exp(1).)

2. Add viscous damping to the system of Problem 1. Use a damping coefficient c = 2 and again solve for the motion resulting from the same initial conditions. Again study the total energy of the system and compare the behavior to what you found in Problem 1.

3. Show that the trigonometric form a cos(x) + b sin(x) is equivalent to the form A cos(x - φ) if $A = \sqrt{a^2 + b^2}$ and the phase angle φ satisfies the equations cos(φ) = a/A and sin(φ) = b/A.

4. Take this opportunity to investigate the physical phenomenon of *resonance* wherein a forced damped oscillator can cause the system to vibrate with unusually large amplitude. Study the system with mass m = 1, damping coefficient c = 2, and spring constant k = 10. Drive this system with the force cos(ω t) and find the steady state solution. The general solution is the sum of the homogeneous solution and a particular solution. The homogeneous solution will contain negative exponentials and does not survive. Hence, it is the particular solution that persists into the steady state. Look for this steady state solution by using the method of **Undetermined Coefficients** and a particular solution of the form yp = a cos(ω t) + b sin(ω t). After solving for the coefficients a and b, put the solution into the form A cos(ω t - φ) via the result of Problem 3. The amplitude A will contain the information about how the magnitude of the steady state solution depends on the driving frequency ω. Plot the amplitude A as a function of ω, and then solve analytically for the exact value of ω that causes A to be a maximum. This value of ω is called the *resonant frequency* and is the frequency at which the driving force generates the largest response from the system.

5. Repeat the study of Problem 4 with the damping coefficient taken to be the parameter c. This means the amplitude A will now be a function of both ω and c, and the resonant frequency computed will exhibit its dependence on the damping coefficient. Plot a graph of the curves that show the amplitude A as a function of the driving frequency ω when c = $\frac{1}{2}$, 1, $\frac{3}{2}$, 2, $\frac{5}{2}$, 3. Also, graph the surface A(ω ,c) over a suitable region in the ωc-plane.

6. Repeat Problem 4 for the system whose mass is m, damping coefficient is c, and whose

spring constant is k. The formula computed for the resonant frequency will now be a function of the three system parameters m, c, and k.

7.   Show that the Euler equation $x^2 y''(x) + 3 x y'(x) + 2 y(x) = 0$ becomes an equation with constant coefficients under the change of variables $x = e^t$. Solve the resulting differential equation and use this solution to write the solution to the original differential equation.

8.   Instead of the *Initial Value Problem* wherein the values $y(t)$ and $y'(t)$ are prescribed at the starting time $t = t_0$, we study the *Boundary Value Problem* for which values of $y(x)$ are prescribed at the boundaries $x = a$ and $x = b$ of some interval $a \leq x \leq b$. Maple's **dsolve** command will solve some boundary value problems, but as long as Maple delivers a general solution of a differential equation, an associated boundary value problem is reducible to algebraic equations in the constants _C1 and _C2 in the general solution.

Consider the differential equation $y''(x) + 2 y'(x) + 10 y(x) = \cos(x)$, with the boundary conditions $y(0) = 0$, $y(1) = 1$. Solve this directly with Maple's **dsolve** command and then solve it by first finding the general solution and imposing the boundary conditions to determine the unknown constants in the solution. Plot the solution on the interval $0 \leq x \leq 1$. Find the coordinates of the maximum point on the graph of this solution.

9.   Solve the boundary value problem whose governing differential equation is the one in Problem 8, but take the boundary conditions as $y(0) = 0$, $y(1) = s$. Study the dependence of the solution on the value of s by plotting on the same axes solution curves corresponding to $s = -5, -4, ..., 5$. Can you determine, as a function of s, the location of the maximum points on these curves?

10.   Study the dependence of the boundary value problem in Problem 8 on the nature of the boundary conditions. For example, take the boundary conditions as $y(0) = 0$, $y'(1) = 1$. Solve and graph the solution. Then take the second boundary condition as $y'(1) = s$ and repeat the study of Problem 9. In addition, how does the value of $y(1)$ depend on s?

11.   Two masses m_1 and m_2 are connected to three springs, S_1, S_2, and S_3, with spring constants k_1, k_2, and k_3 respectively, so that spring S_2 is between m_1 and m_2, S_1 is on the left of m_1, and S_3 is on the right of m_2. The ensemble sits on a frictionless table, and the left end of spring S_1 and the right end of spring S_3 are fixed to posts sticking up from the table at right angles; the posts are far enough apart to cause the three springs to stretch. The equilibrium positions of the two masses are marked, and displacements left or right from these positions are denoted by $x_1(t)$ and $x_2(t)$, respectively. The differential equations of motion for these masses are given by $m_1 x_1''(t) = -k_1 x_1 + k_2(x_2 - x_1)$ and $m_2 x_2''(t) = -k_3 x_2 - k_2(x_2 - x_1)$. Study the motion of the two masses by solving the system of differential equations and graphing $x_1(t)$ and $x_2(t)$. In addition, examine the energies E_1 and E_2 of masses m_1 and m_2, respectively, to see how this energy is transferred back and forth between the masses. This study is best carried out by examining individual cases. Several representative cases are included below.

(a) $m_1 = 1$, $m_2 = 1$, $k_1 = 1$, $k_2 = 1$, $k_3 = 1$

(b) $m_1 = 1$, $m_2 = 3$, $k_1 = 1$, $k_2 = 1$, $k_3 = 1$

(c) $m_1 = 1$, $m_2 = 1$, $k_1 = 1$, $k_2 = 3$, $k_3 = 1$

(d) $m_1 = 1$, $m_2 = 1$, $k_1 = 3$, $k_2 = 1$, $k_3 = 1$

(e) $m_1 = 1$, $m_2 = 3$, $k_1 = 1$, $k_2 = 3$, $k_3 = 1$

(f) $m_1 = 1$, $m_2 = 3$, $k_1 = 3$, $k_2 = 1$, $k_3 = 1$

(g) $m_1 = 1$, $m_2 = 3$, $k_1 = 1$, $k_2 = 1$, $k_3 = 3$.

12. For the system described in Problem 11, include a damping term acting just on mass m_1. Let the coefficient of damping be the as-yet unspecified value c. Also, drive mass m_1 with a forcing term $F \sin(\omega t)$. Show that if $k_2 = m_2\omega^2$ then the steady state solution for $x_1(t)$ approaches 0. This would mean that by attaching extra springs and a mass m_2 to mass m_1 it would be possible to guard against motion of mass m_1, even in the event that mass m_1 is being driven!

REFERENCES

1. Bruce W. Char, Keith O. Geddes, Gaston H. Gonnet, Benton L. Leong, Michael B. Monagan, Stephen M. Watt. *First Leaves: A Tutorial Introduction to Maple V*, Springer-Verlag, 1992.

2. ————. *Maple V Language Reference Manual*, Springer-Verlag, 1991.

3. ————. *Maple V Library Reference Manual*, Springer-Verlag, 1991.

4. Wade Ellis, Gene Johnson, Ed Lodi, Dan Schwalbe. *Maple V Flight Manual*, Brooks/Cole, 1992.

5. Arthur Engel. *Exploring Mathematics with your Computer*, The Mathematical Association of America New Mathematical Library, 1993.

6. Keith O. Geddes, Stephen R. Czapor, George Labahn. *Algorithms for Computer Algebra*, Kluwer Academic Publishers, 1992.

7. André Heck. *Introduction to Maple*, Springer-Verlag, 1993.

8. Robert J. Lopez. *Maple via Calculus: A Tutorial Approach*, Birkhäuser, Boston, 1994.

9. *Maple V Release 3 for the Macintosh: Getting Started*, Waterloo Maple Software, 1994.

10. *Maple V Release 3 Notes*, Waterloo Maple Software, 1994.

11. *The Maple Technical Newsletter*, published Spring and Fall, Birkhäuser, Boston.

INDEX

abs, 1
adaptive procedure, 163
adaptive quadrature, 147
additionally, 165, 173
airplane hangar, 304
allvalues, 39, 45, 212, 279
alternating series, 179
Alternating Series Test, 179, 188
animate, 19, 29, 70, 209
animation problems, 37, 79, 83, 105,
    145, 163, 189, 205, 217, 218, 237,
    238, 239, 252, 288, 304
antiderivative, 85, 86, 103
arc length, 112, 113, 117
arcsin, 137
arcsinh, 119
arctan, 39
area, 85, 92, 101, 102, 103, 104, 105,
    165, 170
assign, 1, 12
assignment operator, 19
assume, 119, 121, 173
asymptote, 88, 89
banana bread, 133
Bezier Curve, 224
bicycle, 76, 81, 223
camera, 73, 145
case sensitive, 47
Cauchy Principal Value, 165, 170, 171
Chain Rule, 56, 283
changvar, 147, 153, 158
characteristic equation, 325
coeff, 197, 203
collect, 177, 261, 321
combine, 197, 201, 332
comet, 78, 205
completesquare, 197, 198, 201, 227
concatenation, 62
constrained, 11
convergence, 168
convert, 119, 126, 139, 140, 147,
    155, 177, 182

cos, 1
contourplot, 261, 262, 263
crossprod, 227, 229, 230
curl, 307
curvature, 245, 247, 251, 255
cylinderplot, 227, 232
D, 39, 42, 137
damped, 168, 332
denom, 85, 88
derivative, 39, 41, 42, 261
det, 227
diff, 39
differential equations, 321
digitizer, 11
directional derivative, 273
display, 19, 70, 71, 73, 208, 209
display3d, 234
diverge, 307
Divergence Theorem, 313, 319
dotprod, 227, 230, 308
Doubleint, 289, 292
doughnut, 306
dsolve, 119, 321
Enter, 2
erase, 1, 13
error, 150, 151, 152, 159
Euler's formula, 325
evalc, 59, 61
evalf, 1, 61, 147, 149
evalm, 272
exp, 107, 119
expand, 1, 86
explicit, 11, 59
factor, 1
feedback control, 157
Fibonacci, 193, 194
fonts, 223
for...do, 19, 27, 56, 57, 62, 73, 105,
    122, 123, 152, 156, 199, 264
Fourier series, 183
floating point, 3, 9, 97, 124
fsolve, 1, 71

function, 1, 4, 5, 19
Fundamental Theorem of Calculus, 85, 96, 102, 147, 159
galaxy, 206
golf ball, 306
grad, 261, 273
Green's Theorem, 307
half-life, 134
harmonic series, 186
headlight, 206
Help Browser, 8
histogram, 289
hula hoop, 223
identity, 156
implicit, 11, 22, 23, 47, 59
implicitplot, 11, 19, 45, 48,200
implicitplot3d, 261, 270
improper integral, 165
inert, 85
infinite series, 177
insequence=true, 19, 30, 70, 209
int, 85, 86, 87, 147
Int, 85, 86, 147, 292
Integral Test, 187, 192
integrand, 86, 87
integration, 85, 87
intparts, 147, 154
inverse function, 119, 120, 137, 138
isolate, 59, 71, 213
Kant, 227
Lagrange Multiplier, 278, 288
Laplace transform, 167
leftbox, 85, 93
leftsum, 85, 93, 291
level curves, 262
lhs, 1, 10
limit, 19, 27, 29
Limit, 19, 28
linalg, 227, 261, 289, 307
Lissajous, 222
ln, 119
log[2], 119, 124
map, 207, 244, 250, 308
Maple Ave., 259
Maple worksheet, 2, 6, 7
matrix, 289
matrixplot, 289, 290
Mean Value Theorem, 145
middlesum, 103, 104, 159
monotone, 122, 138, 177

Monte Carlo, 105, 239
movie, 19, 70, 79, 105, 145, 163, 205, 217, 218, 237, 252, 288
mtaylor, 261, 275, 287
mushroom, 238
Newton Algorithm, 71, 72, 79, 80, 81
Newton's Law of Cooling, 133
NeXT, 2, 6
norm, 241
normal, 147, 156
normal vector, 272
normalize, 227, 230, 245
north polar cap, 296
nose cone, 287
numeric integrator, 147, 149
numpoints, 89
op, 244
osculating circle, 255
parametric equations, 207
parfrac, 147, 155
partial derivatives, 261, 264, 268, 270
partial fractions, 147, 155, 157
partial sum, 179, 185, 186
Pi, 1, 6
plot, 1, 40, 69, 70, 178, 207, 293
plots, 59, 231, 261, 288
plot3d, 227, 231, 263, 289
plot options, 6
pointplot, 234
pointwise, 25
polar coordinates, 207
Principal of Superposition, 325
print, 59, 73
printf, 57
proc, 1
prompt, 2
question mark, 8
random, 105
radar, 207
radon gas, 128
readlib, 59, 140, 213, 261
regions, 2
Rejoice, 67
replot, 14
Return, 2
Riemann integration, 165, 166
rightbox, 107, 111
rightsum, 107, 112, 293
RootOf, 39, 45
save, 8

sec, 85
second derivative test, 64
second partials test, 290
seq, 19, 29, 70, 113, 177, 185, 209
sequence, 177
showtangent, 39, 43, 71
simplify, 1, 61, 68, 147, 289,
  307, 321
simpson, 147, 150, 152, 160, 161,
  162, 163, 164
sin, 1
Snoopy, 224
solve, 1, 19, 65
spacecurve, 227, 231, 243, 297
sphereplot, 227, 233
spherical coordinates, 299, 300
spline, 224, 225
sqrt, 1
stack, 227, 229
Stokes' Theorem, 307, 319
student, 39, 43, 85, 159, 289
  changvar, 147
  completesquare, 197
  combine, 197
  D, 39
  Doubleint, 289
  Int, 85
  integrand, 85
  intparts, 147
  isolate, 59
  leftbox, 85
  leftsum, 85
  Limit, 19
  middlebox, 104
  middlesum, 104
  rightbox, 107
  rightsum, 107
  showtangent, 39
  simpson, 147
  Sum, 85, 92, 96
  trapezoid, 147
  Tripleint, 289
  value, 19
subs, 1, 4, 13, 20
sum, 85, 92
Sum, 85, 92, 96
surface, 261, 263
symbolic, 1, 2, 3, 9, 72, 97, 153
tan, 1
tangent line, 41

tangent plane, 261, 267
taylor, 177, 182, 189, 225, 326
textplot, 59, 71, 73, 293
torus, 306
trapezoid, 147, 149, 160, 162
Tripleint, 289, 298
trisectrix, 214
typesetting, 223
unapply, 1, 40
union, 307
UNIX, 2
value, 19, 85, 86
vector, 227, 228, 308
vector-valued function, 241
volume, 111, 116, 117
Windows, 2
with, 19, 22, 39, 45, 59, 85, 261,
  289, 307
  plots, 59, 261, 288
  linalg, 227, 261, 289, 307
  student, 39, 43, 85, 227, 289
worst case, 151
Zoom, 14